T0228589

UNDERSTANDING AQUACULTURE

Jesse T. Trushenski

5m Publishing

First published 2019

Copyright © Jesse Trushenski 2019

All rights reserved. No part of this publication may be reproduced, stored in a retrieval system, or transmitted, in any form or by any means, electronic, mechanical, photocopying, recording or otherwise, without prior permission of the copyright holder.

Published by
5M Publishing Ltd,
Benchmark House,
8 Smithy Wood Drive,
Sheffield, S35 1QN, UK
Tel: +44 (0) 1234 81 81 80
www.5mpublishing.com

A Catalogue record for this book is available from the British Library

ISBN 9781789180114

Book layout by Servis Filmsetting Ltd, Stockport, Cheshire
Printed and bound in Wales by Gomer Press Ltd
Photos and illustrations as indicated in the text

A special thanks to Dr Matthijs Metselaar DVM PhD MRCVS CertAqV MIFM, for his time and consideration in reviewing this material.

Contents

PART I

Introduction

Chapter 1

What is aquaculture?

Do you enjoy fishing? Do you enjoy seafood? Do you like the idea of your children and your children's children being able to enjoy the same? If so, read on. This book is about aquaculture. It is about what aquaculture can and does provide the human race. It is about what aquaculture does well and where mistakes have been made. It is about informing the public mindset and, with new understanding, embracing aquaculture as a critical element of the future we would like to share with one another. It is, I hope, about understanding aquaculture, fully and appreciatively. First, we must understand aquaculture in the simplest terms – its definition and basic forms.

Aquaculture takes its name in the manner of agriculture, replacing *ager* (Latin for field) with *aqua* (water) and combining it with *cultura* (growing, cultivation) to efficiently describe a multitude of practices (Oxford University Press, 2016). A number of related terms have been invented to try to define the rearing of plants or animals in either fresh or saltwater more narrowly. However, the inherent definition of aquaculture – literally "water cultivation" – applies to all of these endeavors and describes them with a simple elegance the other terms lack. There is a reason the title of this book is *Understanding Aquaculture* and not, for instance, *Understanding Mariculture, Hydroponics, Ocean Ranching, and Fish Farming.* Water is cultivated for a number of reasons, including food production, fisheries restoration, or for ornamental or other purposes. Modern aquaculture is so diverse that it defies even the most basic of categorizations. What is cultured? Throughout the world, hundreds of species of finfish (think salmon, catfish, trout, and so forth), crustaceans (shrimp and crabs), reptiles and amphibians (alligators and frogs), and mollusks (oysters and mussels) are raised. Why are they cultured? Fish and their aquatic brethren are raised as food, for natural resource management purposes, as pets, as experimental models for biomedical research, and so on. How are they cultured? Fish and shellfish are raised in ponds, tanks, cages or net pens, on floating rafts, and in other systems. Farms and hatcheries are located in open water, land-locked locales, and everything in between.

Perhaps because of aquaculture's many forms, the public discourse includes a great deal of mis- and disinformation about what aquaculture is and is not. Is farmed seafood safe to eat? Is wild fish more nutritious than farmed fish? Do hatcheries help or hurt wild fisheries? Don't fish farms pollute the environment? The media breathlessly delivers negative reports on aquaculture but, facts notwithstanding, rarely reports the benefits and importance of farm- and hatchery-reared fish and shellfish. Aquaculture has been practiced for many millennia, yet remains the most misunderstood means of producing food and other goods for human consumption. Roughly one-half of the seafood we eat comes from farms, and yet many consumers remain hesitant, even resistant to buying farmed fish. Aquaculture is compared with both terrestrial agriculture and capture fisheries and has endured considerable, often unfounded, criticism. Whereas terrestrial agriculture is considered a pillar of civilization, the uninformed often dismiss aquaculture as an inferior alternative to fishing. More than ever, consumers want to know where and how their food is produced, and like any agricultural sector, the aquaculture industry has worked to address legitimate questions regarding environmental sustainability, animal welfare, and the sociopolitical dynamics of food security. But with so much conflicting information, it is sometimes difficult to know what is true and what is just a fish story. That is why this book exists and, if my guess is correct, why you have picked it up. *Understanding Aquaculture* will address the common questions and numerous myths surrounding aquaculture and separate fact from fiction. Aquaculture has tremendous promise, but an informed and supportive public is needed to ensure its potential is fully realized. The first part of this book will help familiarize you with aquaculture and how it is practiced throughout the world, and articulate some of the reasons why aquaculture is controversial. The following parts will dive deeper, addressing issues related to the health and safety of farmed fish and shellfish, environmental impacts, and the socioeconomic implications of aquaculture in detail. The concluding part will summarize these subjects and offer thoughts as to why and how we must think about and advocate for sustainable aquaculture.

The precise origins of aquaculture are unknown, likely having been 'discovered' by more than one civilization. However, most historians believe aquaculture has been practiced in China for at least several thousand years. Chinese aquaculture began with cultivation of common carp perhaps as early as 2000–1000 BCE (Rabanal, 1988), the mid to late Bronze Age (Figure 1.1).

By this time, a number of terrestrial animals had already been domesticated, including sheep and goats, cattle, pigs, llamas and alpacas, horses, cats, and, of course, dogs. For context, the first attempts at common carp aquaculture in China likely predated the invention of wheels with spokes, iron-based metals, and the phonetic alphabet. The oldest known monograph describing the cultivation of fish, "The Classic of Fish Culture" was written by the

Figure 1.1: Common carp, the first and most widely cultured fish in the world.

Source: Image from U.S. Fish and Wildlife Service National Image Library, created by Duane Raver.

Chinese politician, Fan Lai,[1] sometime around 475 BCE, meaning that aquaculture had undoubtedly been practiced by Lai, his contemporaries, and their ancestors long before the 5th century BCE.

In Asia and other parts of the world, our long-dead ancestors likely began capturing young fish and fattening them for the table much in the same way that terrestrial livestock were first domesticated. As detailed in the treatise, "History of Aquaculture" (Rabanal, 1988), four scenarios form the basis for how we presume aquaculture arose many millennia ago. According to each of the four complementary theories, observant and enterprising humans began cultivating the water largely for reasons of expediency. The first scenario involves so-called oxbows, the U-shaped lakes that form from the outward-curving bends of rivers as time and erosion work to disconnect them from the main channel. Flood events can rejoin oxbow lakes to their mother rivers, allowing fish and other organisms to move between them. When the waters recede, the corridors of connectivity close and the oxbow is once again separated from the river. For fish stranded in the oxbow, this may be a catastrophe or a stroke of good luck: if the lake dries out completely, all is lost; if not, stranded fish may find refuge from large predators and other dangers of the river, growing large and prolifically. Humans living near the river would almost certainly have known of these rich fisheries and might have begun to artificially recreate the ideal conditions by modifying the embankments of existing oxbows and supplementing the standing stock through periodic introductions of fish.

The second scenario involves seasonal lakes that form in the low-lying areas of tropical regions as flood waters recede following the end of the rainy monsoon season. These lakes could have been similarly improved and supplemented by human communities inhabiting nearby areas. The third scenario extends this concept to coastal regions, where ancient people fishing in pools and coves at low tide may have sought to increase their take by installing traps to keep temporarily stranded fish and shellfish from exiting during the next high tide. In the fourth scenario, humans may have

---

[1] Also spelled Li or Lee, depending on the author.

begun cultivating the water to take advantage of the long-standing premium on fresh seafood. Rather than go fishing in inclement weather or at a moment's notice, servant or peasant classes may have responded to the demands of nobility by holding wild-caught fish in communal water bodies constructed to provide drinking water or defense. Some of the stocked fish would avoid recapture and survive to reproduce and their progeny would be joined by additional fish transferred from natural waters. Each of the theories posits that a population of aquatic organisms is either created or commandeered, and managed – first inadvertently, then through intentional manipulation – to provide greater yields of seafood.

From these humble, perhaps unintentional beginnings, aquaculture grew. By the time of "The Classic of Fish Culture", common carp aquaculture was well-established in China and continued to be refined over the next 1000 years. Until the late 7th century CE, Chinese fish culturists had focused almost exclusively on common carp. This changed in 618 CE, with the seizing of power by the Li family and establishment of the Tang dynasty. The Li family shared its surname with the common name for common carp. After rearing of the new emperor's namesake was prohibited, aquaculturists began raising silver carp, bighead carp, grass carp, and others (Rabanal, 1988). Thus, from imperial vanity came the impetus to establish new husbandry methods and the polyculture (that is, rearing multiple species together) practices that dominate Asian aquaculture to this day. Aquaculture was also developing independently or as a result of immigration in other Asian countries as well as Europe well before the modern era. Interestingly, aquaculture does not appear to have developed in any meaningful way in the Americas, Africa, or Australia until after modern-day introductions of the techniques.

First and foremost, aquaculture was and is an agricultural practice. The Food and Agriculture Organization (FAO) of the United Nations offers a definition that is accordingly agri-centric, defining aquaculture as the "farming of aquatic organisms including fish, mollusks, crustaceans and aquatic plants. Farming implies some sort of intervention in the rearing process to enhance production, such as regular stocking, feeding, protection from predators, and so on. Farming also implies individual or corporate ownership of the stock being cultivated, the planning, development and operation of aquaculture systems, sites, facilities and practices, and the production and transport" (Food and Agriculture Organization, 1988: 37). Until the 1990s, however, aquaculture was a relatively insignificant contributor to the global seafood supplies (Figure 1.2).

To this point, increasing seafood demand was largely met by increasingly industrialized fishing effort that exploited marine stocks with unprecedented efficiency. The oceans' bounty, once thought to be an inexhaustible resource, proved no match for human ingenuity and drive and, under the weight of overfishing and environmental change, fisheries contracted and collapsed one after another. In 1974, only 10% of marine fish stocks were considered overfished (harvested in excess of sustainable catch estimates) and 40% were considered underdeveloped; whereas the majority of these fisheries

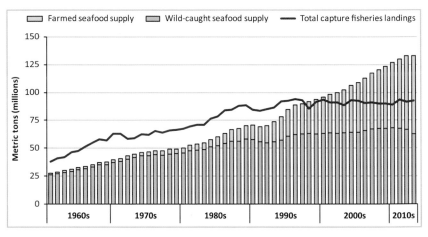

Figure 1.2: Relative contributions of capture fisheries and aquaculture to world seafood supplies, 1961–2013. As fishing fleets grew in size and in technological sophistication, total fisheries landings, including fish and shellfish used for food and industrial purposes, increased steadily through the 1980s. Thereafter, declining populations and implementation of stricter quotas and regulations on fishing effort have stabilized fisheries landings at roughly 90 million metric tons per year. Consequently, the supply of wild-caught seafood changed little during this time period. In response to ever-growing seafood demand and static supplies of wild-caught product, the aquaculture industry has grown dramatically for decades. Today, more than one-half of seafood eaten throughout the world every year is farmed.

Source: Data adapted from Food and Agriculture Organization (2016a, 2017a, 2017b).

are still considered to be fully, but sustainably fished (a little more than 60%), the relationship between under- (slightly under 10%) and overfished (nearly 30%) stocks has essentially reversed over the last 30 to 40 years (Food and Agriculture Organization, 2014). Capture fisheries landings have held at approximately 90 million metric tons per year since the 1990s, about three-quarters of which is consumed by people. Despite static supplies, seafood demand continues to grow, currently topping more than 130 million metric tons. Dubbed the "seafood gap", this represents a shortfall of more than 60 million metric tons every year (Food and Agriculture Organization, 2014). Demand for seafood is primarily driven by human population growth. There are more than 7 billion of us scrabbling about on Earth (Population Fund, 2014), and on average we eat a little more than 19 kg (almost 42 pounds) of seafood a year (Food and Agriculture Organization, 2014). Based on recent population growth rates of a little more than 1% per year and static capture fisheries landings, the 60 million metric ton seafood gap increases by more than 2.5 metric tons every minute. That is how much fish, shrimp, and other assorted seafood we all rely on commercial aquaculture to produce.

Aquaculture is also a natural resource management activity. Even before marine commercial fisheries began to show the signs of over-use, many freshwater recreational fisheries had slumped and staggered into a state of decline. At the end of the 19th century, many freshwater fisheries in North America had deteriorated as a result of overharvest, habitat modification, and other pressures. In the USA, the US Fish Commission[2] was formed in 1871 to determine why fish stocks throughout the country were in decline and, importantly, what could be done about it. One of the Commission's first goals was to produce and stock shad and salmon in regions where these fish were no longer abundant. The public was eager for technological solutions to societal and environmental problems, and fish

---

[2] Both the US Fish and Wildlife Service and the National Marine Fisheries Service trace their origins to the US Fish Commission. They, along with state fisheries agencies, continue to serve in much the same capacity as the US Fish Commission – to pinpoint and resolve the problems of declining fish stocks in the USA.

hatcheries were viewed as modern marvels that would turn depleted waterways into bountiful sources of food and recreation, eliminating the need for fishing quotas. It was believed that hatcheries would compensate for the effects of habitat degradation, overharvest, and the other fishy woes wrought by man. Many US hatcheries were opened during the golden era of dam construction to seed fisheries in newly created reservoirs and to mitigate the loss of once robust riverine fisheries following the construction of hydropower and water control structures. Hatcheries were built and operated throughout North America with an enthusiasm matched only by our confidence in being able to bend the processes of the natural world to the will of a growing populace (Figure 1.3).

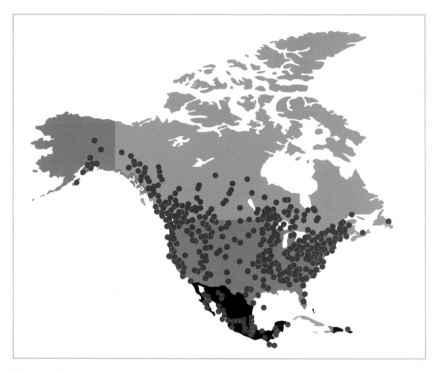

Figure 1.3: Approximate locations of public fish hatcheries engaged in aquaculture for natural resource conservation purposes, including fisheries enhancement and restoration.

Source: Facility and location information sourced from federal and state natural resource agencies in the USA, Canada, and Mexico. See Trushenski et al. (2018a) for additional details. Artist: Jorgen McLeman.

For decades, this thinking persisted, even as many fisheries continued to decline. As time marched on and catches dwindled, it became clear that fish hatcheries, gleaming monuments to efficiency, simply were not living up to their technological promise. "With the passage of time and growing realization that stocking and restrictive regulation were ineffective, fisheries workers began to search for new answers ... The focus of fisheries management broadened from the previous narrow fixation on fish culture to more appropriate, ecologically oriented programs" (Stroud, 1986: 7–8). Today, hatcheries operate very differently than in the early days based on what their products are supposed to do after they are stocked. Thanks to improvements in habitat, changes in harvest regulations, and better use of hatchery fish, many fisheries have been brought back from the brink. Better management, operational vision, and – most importantly – results have kept public aquaculture at the heart of natural resource management (Trushenski et al., 2015). Although this book is primarily focused on commercial aquaculture, it is important to recognize that aquaculture is not exclusively about seafood production.

Aquaculture is a hobby and profession. A goldfish, perhaps the result of a lucky ring toss at the local fair, is a common first pet throughout the world. Roughly one out of ten households in the USA own fish, totaling more than 100 million specimens (American Pet Products Association, 2016). One wonders how many of us respond to the routine password protection question, "What is the name of your first pet?" with some variant of Goldie or Bubbles. Not only are these millions of pet owners practicing aquaculture themselves, it is quite likely that their prized pets came from an aquaculture farm, too. Most of the fish dissected by squeamish seventh-graders in science classes are farm-raised. The less squeamish of those students may grow up to use aquaculture, indirectly, in their work with the National Institutes of Health's Zebrafish Core, part of the Eunice Kennedy Shriver National Institute of Child Health and Human Development (US National Institutes for Health, 2016). These biomedical scientists and other researchers engaged in aquaculture are joined by the roughly 19 million others throughout the world who raise fish and shellfish for a living (Food and Agriculture Organization, 2014).

More than half the seafood we eat comes from farms. Worldwide, farms now produce more seafood than beef. Aquaculture is the most efficient means of turning feed-grade protein into food-grade protein. Farmed seafood is wholesome, nutritious, and essential to global food security. Without hatcheries, many recreational, commercial, and subsistence fisheries would not exist as they do today. Aquaculture is a significant economic driver and supports science and innovation. This book will spend much more time dealing with the myths, but these are the truths of aquaculture.

Chapter 2

The state of aquaculture in the context of global food supply

Although aquaculture has existed for millennia, one could be forgiven for not having noticed it until relatively recently. From 1950 to 1982, global aquaculture production grew ten-fold, from roughly 550,000 metric tons to nearly 5.5 million metric tons. A ten-fold increase in less than 30 years seems rather impressive, until one compares the aquaculture production statistics to capture fisheries landings from the same time period. In 1982, the wild-caught seafood supply topped 48 million metric tons, making the volume of farmed fish and shellfish produced at the time a proverbial drop in the bucket (Food and Agriculture Organization, 2017a).

In the intervening years, capture fisheries landings have not changed appreciably: since the 1990s, in rough numbers, capture fisheries landings have hovered around 90 million metric tons, yielding 60–70 million metric tons of seafood. Seafood demand, on the other hand, has grown steadily, creating a "seafood gap" – the difference between demand and the supply provided by capture fisheries. To close this gap, aquaculture has maintained an impressive upward trajectory for decades: the industry has topped itself every year since 1962, setting a new production record annually for more than 50 years (see Figure 1.2). As a result, aquaculture has gone from producing only 5% of the world's seafood to overtake capture fisheries as the leading source of seafood, producing nearly 74 million metric tons in 2014 (Food and Agriculture Organization, 2016b).

In simple terms, seafood demand is a function of population size and per capita seafood consumption. Increases in both have synergistically driven seafood demand upwards for many decades, but neither population growth nor seafood consumption is homogenous throughout the world. Population size is currently greatest in expansive regions of South Asia, East Asia, and the Pacific, and though growth rates have slowed somewhat in recent years, they remain solidly positive in most of these countries (Figure 2.1). Seafood consumption varies considerably among these Asian and Pacific nations, however, fish and shellfish tend to feature prominently in typical diets (Figure 2.2). In contrast, the countries of Europe and Central Asia, Latin America and the Caribbean, the Middle East and North Africa, and North America represent significantly smaller segments of the global population. In many of these countries, particularly in Western Europe and North America, growth rates are slowing, even reversing. Seafood is an important dietary staple in some of these regions, but consumption is influenced by geography, economics, and cultural norms. Coastal and island nations have direct access to seafood, and stronger tastes and food traditions involving seafood as a result.[1] Despite the common association of certain seafood products (for example, lobster, caviar, high-end sushi) with eating habits of the affluent, seafood can also be an inexpensive source of animal protein. On average, residents of industrialized countries do tend to consume more seafood (26.8 kg/year) than those living elsewhere (12.4–20.0 kg/year); however, these statistics likely underrepresent the contributions of subsistence fishing and aquaculture in developing countries (Food and Agriculture Organization, 2016b).[2] Seafood consumption is

---

[1] Iceland provides a good example of the influence of access and food traditions (Icelanders consume an average of 250 g/capita/day), but nowhere is this more obvious than in the Maldives, where residents of the archipelago nation consume more than 500 g/capita/day.

[2] Seafood consumption is highest in industrialized nations (26.8 kg/year) and lowest in the least-developed countries (12.4 kg/year). Beyond this simplistic comparison of those with the most and least buying power, economic standing does not appear to influence seafood consumption in a direct, linear fashion. Consumers in other developing countries do consume more seafood than those in the least-developed countries (20.0 kg/year), but those in developed, but not yet industrialized countries do not (13.9 kg/year). As consumers acquire

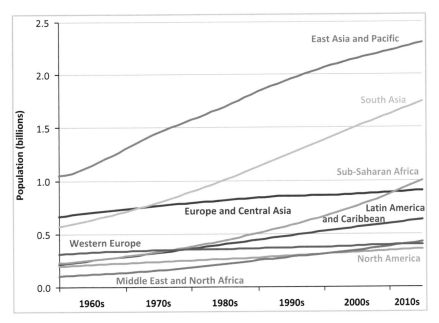

Figure 2.1: World population size (billions) from 1961 to 2013. On average, global population growth rates have declined over this period, from approximately 2.3% annually to 1.3%. However, population size and growth rates vary substantially among regions. Whereas growth rates are flattening in Europe and Central Asia, Western Europe, and North America (even becoming negative in some countries), growth rates are steady or increasing in East Asia and Pacific, South Asia, Latin America and Caribbean, and, especially, Sub-Saharan Africa (World Bank, 2017a).

influenced, at least in part, by whether or not consumers have access to beef, pork, or other types of animal protein that might otherwise be preferred: in places where such meats are too costly, seafood may routinely take their place. Sub-Saharan Africa recently overtook the New World and European nations in terms of population size, and is growing at a rate that is twice the global average. Seafood

greater buying power, there appears to be a transition from eating little animal protein (least-developed countries), to eating more seafood (other developing countries), to eating less seafood and more terrestrial animal proteins (developed countries), to eating more of all types of seafood and meats, including more luxury products (industrialized countries) (Food and Agriculture Organization, 2016a).

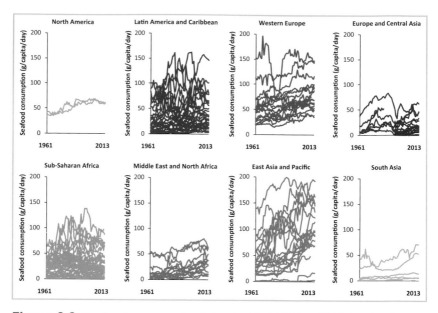

Figure 2.2: Seafood consumption (g/capita/day) 1961 to 2013. Seafood consumption is generally increasing throughout the world, but there is considerable regional variation in current consumption and trends. Although there is considerable variation among countries in Latin America and Caribbean, Western Europe, East Asia and Pacific regions, and Sub-Saharan Africa regions, seafood tends to be a more important dietary element in these places than in the North America, Middle East and North Africa, Europe and Central Asia, and South Asia regions (Food and Agriculture Organization, 2016a). Variation within regions is driven by a number of factors, including geography (consumption tends to be higher in coastal or island nations), economic standing (seafood consumption is sometimes lower when consumers have access to other animal proteins), and cultural norms (seafood typically remains high in cultures with strong food traditions and preferences involving fish and shellfish). Regional boundaries generally follow those used by the World Bank.

consumption varies among these African nations, too, based on the relative ease of accessing seafood versus other animal proteins and the other aforementioned socioeconomic factors.

In broad terms, seafood consumption is high in regions with the largest and fastest growing populations in Asia and Africa, including a large number of developing countries. Global demand for seafood could be expected to grow steadily under these conditions alone, but seafood consumption is also increasing in the more static or contracting population segments in Europe, North America, and Latin America. Whether driven by the desire for more animal protein or the specific nutrients found in seafood, such as certain vitamins, minerals, or "omega-3" fatty acids, seafood intake is on the rise. In 2013, global seafood consumption averaged approximately 19 kg per person, more than double what it was at the beginning of the 1960s (Figure 2.3). Of course, this generalized trend does not reflect dramatic differences in seafood consumption patterns from country to country. Even if one ignores the uniquely high seafood consumption rates of Icelanders and Maldivians, intake in the top seafood-consuming nations (~40–70 kg/capita/year) is 10- to 700-fold higher than in the countries with the lowest consumption of fish and shellfish (less than 4 kg/capita/year) (Figure 2.3).

Seafood consumption patterns have also changed with respect to what types of fish and shellfish are produced and consumed (Figure 2.4). Fifty years ago, 95% of the seafood supply came from wild catches. Consequently, marine fish represented nearly 70% of the seafood supply, with mostly wild and mostly marine crustaceans, mollusks, and cephalopods providing an additional 15%. Freshwater fish and other aquatic animals rounded out the remaining 15% of production. Five decades later, the relationship between marine- and freshwater-origin seafood has shifted, with marine fish, freshwater fish, and the remaining categories now representing about one-third of the supply each.

The supply of cephalopods has changed relatively little over the past 50 years. The catch is dominated by species such as the Argentine shortfin squid and jumbo flying squid or Humboldt squid. These and other various squid, cuttlefishes, and octopi consumed throughout the world are harvested from the wild. Despite recent advances and commercial expansion in marine aquaculture, marine fish are still mostly wild-caught, though the total supply is smaller now than in the past. The catch is dominated by white-fleshed fish, such as Alaskan pollock or Walleye pollock and Atlantic cod, but also includes sizable volumes of skipjack tuna, yellowfin tuna, and other pink-fleshed fish. The crustacean seafood supply includes a number of wild-caught and farmed species. Catches of *Acetes japonicas,* a small krill-like species used along with other members of its genus to make akiami shrimp

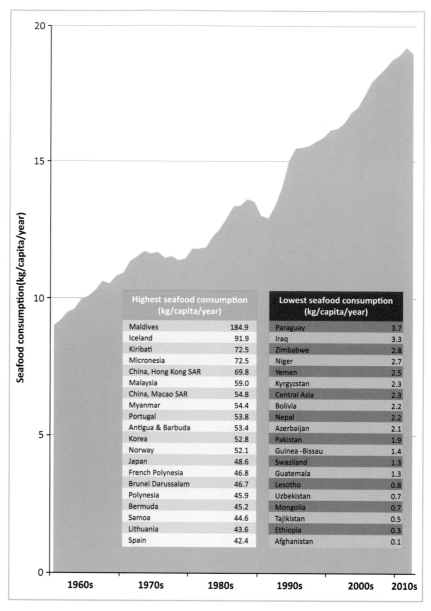

Figure 2.3: Global seafood consumption (kg/capita/year) from 1961 to 2013. Worldwide, annual seafood consumption has more than doubled since the 1960s, but this general trend belies wide national variation in consumption patterns. Countries representing the top 10% and bottom 10% by annual per capita seafood consumption are highlighted (Food and Agriculture Organization, 2016a).

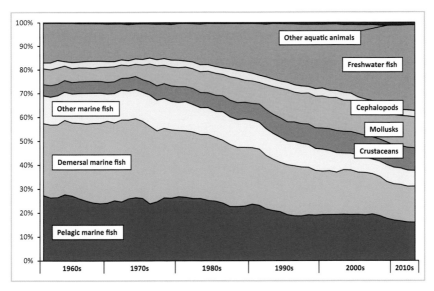

Figure 2.4: Relative contributions of various species groupings to global seafood supply from 1961 to 2013. Historically, the seafood supply was dominated by wild catches of marine fish and shellfish, but the relative contributions of marine versus freshwater organisms, vertebrates versus invertebrates, and so on, have shifted because of dramatic increases in aquaculture production over the past 50 years (Food and Agriculture Organization, 2016a).

paste, come from the largest shrimp or prawn fishery in the world. Fisheries for blue crabs and snow or spider crabs are also quite large. Giant tiger prawn and Pacific whiteleg shrimp are the most important farmed crustaceans, and collectively represent a full third of the crustacean seafood supply. Like crustaceans, mollusks are both fished and farmed, and supplies have expanded somewhat in recent years because of increased farming activity. Aquaculture provides much of the production volume in the form of Japanese carpet shell or Manila clam, blue mussel, and Pacific cupped oyster and other members of its genus. Although freshwater capture fisheries are regionally important, the major sources of freshwater fish are all farmed. In fact, the freshwater fish sector has grown dramatically in recent years, almost exclusively because of increases in aquaculture. Production is dominated by Asian carps, such as bighead carp, silver carp, grass carp, and common carp, and Indian carps, such as catla and rohu.

Various farmed tilapias, most notably Nile tilapia and its hybrids, are also major contributors to the freshwater fish supply. Important diadromous fishes include a number of species that are predominantly farmed, such as milkfish, Atlantic salmon, and rainbow trout/ steelhead (Food and Agriculture Organization, 2017b).

Looking to the future, it is clear that seafood demand will continue to grow, but capture fisheries will not. Although many fisheries were up in 2014 relative to the previous few years, capture fisheries landings have not increased appreciably for 30 years. Indeed, some have argued that over-reporting by some nations has masked declining trends in marine harvest (Watson and Pauly, 2001). Regardless of whether one believes the volume of landings is stable or declining, it is undoubtedly true that the nature of the fishing industry and the composition of the catch has undergone substantial evolution over the past five decades. During this time, the fishing fleet has become industrialized throughout much of the world: small, independently owned and operated vessels have been displaced by corporately held leviathan factory ships; traditional ecological knowledge of fish movements has been replaced by sophisticated tracking systems; and modern ship-building, navigation systems, and mechanized, ruthlessly efficient fishing gear have rendered the limits of human strength and comfort quaint in the exploration and exploitation of the seas.[3] For many decades, the species composition of marine catches has shifted from large, long-lived species at or near the top of their respective food webs, to smaller, short-lived species occupying lower trophic levels in their ecosystems. Described as "fishing down the marine food web", the shift from mostly predatory species like tunas and billfishes to an increasing number of filter-feeding species of fish and invertebrates[4] indicates declining abundance of the larger, more valuable species (Paul et al., 1998).

---

[3] In describing the Alaskan Pollack fishery, Becky Mansfield notes that one-half of the catch – approximately 1.5 million metric tons – is caught by 120 vessels, mostly owned by or otherwise obligated to a handful of multinational seafood companies (Mansfield, 2011).
[4] New fisheries for jellyfish might be considered the most extreme example of fishing to the bottom of a food web. Fisheries scientists are uncertain as to

Though it lacked a pithy phrase until the late 1990s, this phenomenon and its potential consequences were already well-known to the fisheries science and policy communities. Commissioned by then-U.S. President Jimmy Carter, the "Global 2000 Report to the President of the U.S." (Barney, 1980: 313) predicted the following of global fisheries:

> Over the last three to four decades, intensive fishing activity has produced a shift in the species composition of the global catch away from the traditionally preferred species towards species at lower trophic levels and of less economic value … [D]emand for seafood will increase, encouraging still more fishing activity … Future gross catch statistics therefore may show a constant or increasing yield, but the catch will be composed of progressively less traditional products. Advances in fishing and processing technologies, by helping the gross catch figures to remain high, will effectively conceal the degree to which overfishing is undermining the utility and value of the world catch.

Indeed, this is consistent with the history of many fisheries, as chronicled by such authors as Paul Greenberg (*Four Fish: The Future of the Last Wild Food*, 2010), Mark Kurlansky (*Cod: A Biography of the Fish that Changed the World*, 1997) and others: a fishery is developed; becomes lucrative; removes the largest individuals in a population, then the next largest, and so on; collapses under the pressures of added fishing effort and increasingly efficient exploitation; and is abandoned for the next fishery. Today, regulations controlling when and how fishing is done and how many fish can be harvested are put in place to begin the slow process of rebuilding collapsed fisheries and to prevent others from meeting the same fate. Some of these interventions have been successful and there are a good number of sustainably managed fisheries throughout the

---

whether these new and growing fisheries represent a predictable response to growing demand in Asian markets, or an artifact of increasing jellyfish abundance as a result of environmental degradation (Food and Agriculture Organization, 2016b).

world, but many would argue that the apparent stability of capture fisheries landings belies a crisis slowly unfolding in the manner foretold more than 35 years ago in the "Global 2000" report. Catches have been stable, but marine fish biomass has continued to shift toward smaller fish and lower trophic levels, albeit at slower rates than during the 1970s-1980s heyday of industrialized exploitation of the seas (Christensen et al., 2014; Pauly et al., 2005).

The intelligentsia of fisheries science continue to debate the data and what they imply for wild fish stocks, the relative strengths and weaknesses of existing regulatory frameworks and policies, and what is to be done about any of it. This discourse is critical to the future of global fisheries, but details aside, it is becoming increasingly clear that we will not get any more food from the seas in the future – if anything, we may get less. If progress to slow the decline of capture fisheries is to continue, the only solution to the seafood gap is growth in global aquaculture.

Chapter 3

Aquaculture from place to place

Aquaculture is diverse. Those few words seem inadequate in describing the myriad ways people are raising aquatic organisms – many hundreds of species – throughout the world. Critics often present aquaculture as a largely homogenous endeavor, suggesting that issues like the use of wild fish in salmon feeding or modification of coastal habitat to construct shrimp ponds are systematic problems that affect all sectors equally. In truth, salmon and shrimp farming are sizable industries, but represent comparatively small segments of global aquaculture production volume and value (Figure 3.1). Whatever the historical or current issues facing salmon and shrimp production, they can hardly be said to be representative of aquaculture as a whole.

What aquatic organisms are produced in aquaculture and how they are reared varies considerably from place to place. Beyond taxonomy, there are a number of ways the staggering diversity can be broadly parsed: extensive versus intensive aquaculture, commodity versus specialty products, high-value versus low-value species, herbivorous versus omnivorous versus carnivorous animals, fed versus unfed species, and so on. It may be impossible to fully *know* aquaculture as an enterprise, but understanding how all of this diversity is categorized can give one an appreciation for the varied ways aquaculture is practiced throughout the world.

Extensive and intensive aquaculture occur on a spectrum, ranging from methods separated from natural productivity by the barest of interventions to wholesale abandonment of nature and the

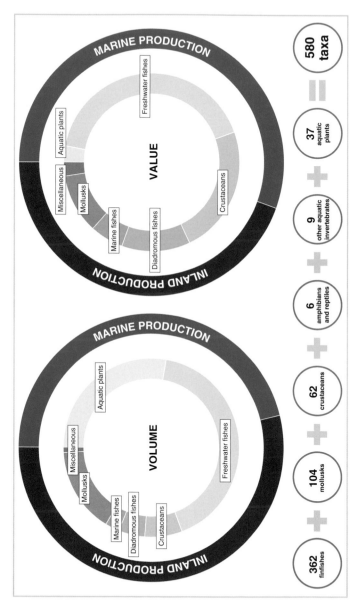

Figure 3.1: The aquaculture industry is exceptionally diverse, in terms of what species and how they are raised. Nearly 600 types of aquatic plants and animals are raised throughout the world, in both marine and inland waters (Food and Agriculture Organization, 2016b). For many, aquaculture conjures thoughts of farmed shrimp and salmon, but these two sectors accounted for less than 12% of global production volume (in metric tons) and 36% of production value (in US$) in 2015 (Food and Agriculture Organization, 2017a).

creation of artificial ecosystems in which to rear aquatic organisms (Figure 3.2). Aquaculture's origins are in allowing aquatic organisms to exist in a semi-controlled environment. Our ancestors, noting the numbers and size of fish left behind in oxbow lakes during dry seasons, likely intervened to make such strandings more common-place. Protected from predators and competitors that would other-wise curb their productivity, the oxbow-bound fish (and later, their human patrons) would have benefited from the arrangement. This is aquaculture at its most elemental and extensive: only by exerting some control over the species composition of the oxbow lake does man cultivate the water. At some point, man discovered that fertiliz-ing the water, perhaps with manure or other organic wastes, stimu-lated the bottom of the food chain and provided his captive fish with a bit more to eat. Soon after, he likely learned – perhaps the hard way – that such additions and increases in biomass could cause an

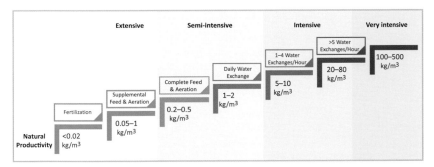

Figure 3.2: Depending on the number of inputs and the level of control the culturist has over the cultivated organisms and rearing environment, aquaculture can be extensive (few inputs, low rearing densities), intensive (many inputs, high rearing densities) or somewhere in between. Moving from left to right, carrying capacity is artificially increased by providing oxygen and food and removing wastes to overcome the biological limitations that would otherwise curtail productivity. One drawback to intensive aquaculture is its inherent risk: unless all life support systems are maintained continuously, the constraints of oxygen availability and waste accumulation quickly reassert themselves. As one culturist puts it, raising fish intensively is like driving fast on a curvy road: one small miscalculation and the vehicle is hurtling toward the ditch. Like driving at more reasonable speeds, extensive aquaculture is more forgiving and usually allows for mistakes to be corrected before catastrophe ensues.

Source: Adapted from Piper et al. (1986).

imbalance and the entire crop could be lost. As man's understanding of his pond's dynamics and potential grew more sophisticated, he may have begun to dream of new ponds, built according to his ideals of bountiful crops and easy harvest. Millennia have since passed, and the limitations encountered by proto-aquaculturists can now be readily overcome through any number of technological innovations to increase oxygen availability, process fish wastes, and increase rearing densities without causing catastrophic failure. Intensive aquaculture involves taking responsibility for what was once left to Mother Nature, using technology and management to increase the carrying capacity leaps and bounds above what natural conditions would otherwise support. By divorcing the rearing of aquatic organisms from the vagaries of the aquatic environment, the most intensive, contained aquaculture systems can achieve fish densities 10,000 times greater than that possible with extensive methods. Although it would be easy to associate intensification with modernization and assume that most aquafarms are testaments to intensification, this would be incorrect. In fact, aquaculture practices throughout the world are mostly extensive or semi-intensive. The vast majority of fish and shellfish are cultivated in carefully managed ponds or closely monitored pens (Figure 3.3).

Figure 3.3: In Argentina, a fish farmer feeds golden dorado cultured extensively in an earthen pond. Although golden dorado are a highly prized, top-dollar fish, the farmers provide them with little other than feed and, when necessary, supplemental aeration to maintain oxygen levels. In Canada, Atlantic salmon are raised in coastal net pens. Despite modern engineering and monitoring tools, this method of raising salmon is also extensive, given that the only input is feed.

Source: Photos by the author.

This is not to say that these production systems do not bear the marks of modernization – such as probes that monitor dissolved oxygen levels and notify farmers of declining pond conditions via text message or underwater video cameras used to watch and gauge feeding behavior in sea cages – only that the most intensive forms of aquaculture are still relatively rare. Mother Nature may be cruel, but her wages are low: extensive aquafarmers relying on natural ecosystems to care for their livestock pay little or nothing for these services. Intensification promises greater harvests, but also increasing costs of production and risk. Throughout the world, aquaculture is mostly a compromise between the relative ease and low cost of extensive methods and the productivity and predictability of intensive methods (Figure 3.4).

Only a fraction of the diversity of the natural world is reflected in aquaculture, yet the industry is strikingly more taxonomically varied than any other form of animal husbandry.[1] Terrestrial animal agriculture boasts hundreds and hundreds of domesticated strains and lines, but relatively few true species. Poultry is the second-most taxonomically diverse form of animal agriculture, but is dominated by chickens, turkeys, ducks, geese, quail, and a small number of less-common avian species raised for their meat and eggs. Of the 150 wild and domesticated ruminants, milk and meat production is dominated by a few species of cattle, goat, and sheep. Pork is the least diverse of the major livestock commodity groups, coming from a single species. Embedded in the taxonomic diversity of aquaculture is wide-ranging variation in environmental preferences. There are fish and shellfish that are adapted to survive in water temperatures as warm as 33°C or as cold as 1°C.[2] There

---

[1] Taxonomists have described more than 33,000 species of fish, roughly 8000 species of bivalves, and more than 70,000 species of crustaceans, of which the vast majority are aquatic. These multitudes include the 528 finfish and shellfish that are produced in aquaculture, but not the handful of other aquatic animals, plants, and algae that are cultivated throughout the world.

[2] Tilapias, cultured throughout the world, but native to Sub-Saharan Africa, can tolerate temperatures as high as 33°C. Siberian sturgeon, cultivated in their native Russia, can withstand temperatures approaching freezing at 1°C (Froese and Pauly, 2017). For those who are thermally challenged, this range represents average high temperatures in Chicago, USA in July vs. February.

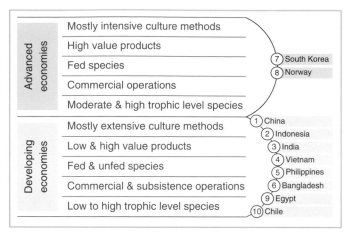

Figure 3.4: The top ten aquaculture producing countries vary in terms of the species reared, purpose and methods of production, and intended markets (Food and Agriculture Organization, 2016b). As one might expect, all of the top aquaculture producing countries are possessed of expansive coastlines or other significant water resources, and this partly drives what comprises aquaculture in these places. For example, Norway's 63,000 miles of coastline and cold, deep fjords are near perfectly suited to salmon culture in net pens, whereas China's inland freshwater resources and immense agricultural lands (>50% of the country's land mass) are best suited to land-based culture of carp. The form and function of aquaculture across the world is also influenced by economic standing. South Korea and Norway are considered advanced, high income economies, whereas the remaining eight countries are considered developing economies. There is, of course, considerable variation among these eight nations: Chile is a high income country and China is considered an upper middle income country, but the others are all considered lower middle income nations (International Monetary Fund, 2017; World Bank, 2017b). Factors such as technological innovation, industrial capacity, buying power, labor costs, and competition for land use foster intensive aquaculture in advanced economies, with an emphasis on high-value species. Conversely, extensive methods dominate aquaculture in developing economies, particularly those with low labor costs and easy access to land and water resources. Although subsistence aquaculture and cultivation of products for domestic consumption is common in countries with developing economies, many such nations increasingly emphasize high-value aquaculture products intended for export.

are species adapted to live in fresh, brackish, or saltwater, as well as unique species that can adjust to virtually any salinity. Most aquatic animals do best in neutral or slightly alkaline waters, but there are those that tolerate, even prefer conditions that are a bit more acidic or basic.[3] Where there is water, there is aquatic life adapted to it. And where there are entrepreneurial men and women, that aquatic life is being cultivated.

Most people would think that cattle, chickens, and pigs have nothing in common,[4] but animal nutritionists are not most people. Where others see hair and hooves, feathers and fur, and countless other ways to differentiate animals, nutritionists see only monogastrics (poultry and swine) and ruminants (cattle, sheep, goats). Monogastric animals have, as their name implies, one stomach,[5] ruminants have multiple stomachs with which to extract nutrients from their food. In terms of how they break down food, domesticated animals boil down to just these two strategies. The same can be said of what and how they eat: terrestrial livestock are middling creatures – not quite the bottom of the food chain, but nowhere near the top. They are herbivores and omnivores, occasionally hunting, but mostly grazing their way through life. Of the tens of thousands of described piscine taxa, there are no ruminants. There are a few agastric fish, lacking stomachs altogether, but most fish are monogastric. This they have in common with chickens, pigs, and their one-stomached domesticated brethren. But there are other monogastrics, found beyond the barnyard. Lions. Tigers. Bears.[6] All are monogastrics. There are some herbivorous and omnivorous fish – even some detritivores that subsist principally on dead and decaying plant and animal matter – but there are many more carnivores, and they have just as much, if not more in common with these charismatic terrestrial predators (Figure 3.5). Tens of thousands of aquatic species have co-evolved with tens of

---

[3] Black pacu, a popular food fish throughout Central and South America, can tolerate acidic waters with a pH as low as 5 (Froese and Pauly, 2017). Common substances with a pH of 5 include black coffee, pickle juice, and some soft drinks. Lahontan cutthroat trout, in contrast, are adapted to withstand waters as basic as pH 9 or greater. For reference, chemical solutions used to curl or 'perm' hair usually have a pH of 8.5–9.5.

[4] Except, perhaps films: with *Charlotte's Web*, *Babe*, *Chicken Run*, *Barnyard*, and countless others, childhood is virtually overrun with farm animals.

[5] Readers will find this arrangement familiar: humans are also monogastric.

[6] Oh, my!

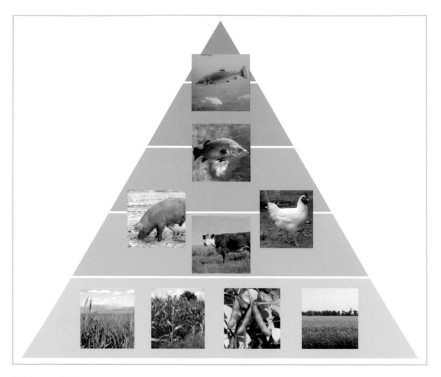

Figure 3.5: Terrestrial agriculture is populated primarily by species at the bottom of the food chain. Grains, cereals, oilseeds, and other plant crops are primary producers – like other photosynthesizers, they are assigned a trophic level of 1. Primary consumers or herbivores like cattle, sheep, and other domesticated grazers occupy trophic level 2. Trophic level 3 is home to secondary consumers – carnivores that eat herbivores. Pigs and chickens are omnivorous, and their diet of plant and animal material places them somewhere between trophic level 2 and 3. Trophic level 4 represents tertiary consumers – carnivores that eat other carnivores. None of the commonly raised terrestrial livestock fit this category, but many aquatic livestock do. Apex predators themselves have no natural predators and consume various other carnivores. As is often said, it is lonely at the top: apex predators represent an enormous amount of energy inefficiently accumulated up through the food chain and most ecosystems can only support a few of these rare animals.

Sources: Hereford cow / Author: Michael MacNeil, U.S. Department of Agriculture / Wikimedia Commons / Public Domain; broiler chicken © User: Cros2519 / Wikimedia Commons / CC-BY-SA-4.0/GFDL; sow / Author: Scott Bauer, U.S. Department of Agriculture / Wikimedia Commons / Public Domain; wheat / User: H20 / Wikimedia Commons / Public Domain; Atlantic salmon © User: Atluxity / Wikimedia Commons / CC-BY-SA-2.5; soybeans and corn © Author H. Zell / Wikimedia Commons / CC-BY-SA-3.0/GFDL; canola © Author John O'Neill / Wikimedia Commons / CC-BY-SA-3.0/GFDL; and barramundi © Author Mitch Ames / Wikimedia Commons / CC-BY-SA-4.0.

thousands more types of forage and prey. Although the diets of most cultured fish and shellfish are less varied than their wild or feral counterparts, the complexity of satisfying their nutritional demands remains. They are what they eat, and they and their diets are diverse.

Feeding practices are another means by which aquaculture operations are categorized, incorporating differences in the cultured species and how intensively they are reared. Nearly 74 million metric tons of finfish and shellfish are raised throughout the world each year, but nearly one-third are not fed, at least not directly. So-called "unfed" species are filter-feeding mollusks and finfish that sieve an exclusively planktonic diet from the waters they inhabit. Farmers rely on natural forage to feed unfed livestock, but in some cases they may harness and whip natural productivity into a frenzy, fertilizing ponds to stimulate phyto- and zooplankton "blooms" that feed their fish. Fed species are offered nutrition they are intended to consume directly. In less intensive culture systems where fish have access to some natural forage, direct feeding may be supplemental and only intended to boost growth rates and harvest volumes. When direct feeding is the primary or sole source of nutrition, nutritionally complete feeds must be used, typically in the form of industrially compounded, commercially available aquafeeds. Aquaculture has been trending toward increasing intensification and use of complete feeds has become more prevalent, but clams, oysters, mussels, filter-feeding carps, and other unfed species account for more than 30% of global aquaculture production. Fed aquaculture production has expanded more rapidly than unfed aquaculture in recent years, but the unfed sector is currently underdeveloped in Africa and Latin America and may grow to address unmet food security challenges in these regions (Food and Agriculture Organization, 2016b).

The top ten aquaculture producers in the world – China, Indonesia, India, Vietnam, Philippines, Bangladesh, South Korea, Norway, Egypt, and Chile – are collectively responsible for more than 90% of the volume of farmed fish, crustaceans, and other seafood produced annually and more than 80% of its value. Though one cannot safely say that these nations represent aquaculture as a whole, by and large, they do provide a reasonable snapshot of aqua-culture from place to place.

Even within the list of top ten, juggernaut nations, China is overwhelmingly dominant in the aquaculture sector. From 1993 to 2003, the Chinese aquaculture industry grew threefold (from <10 million metric tons to >30 million metric tons), and has more than doubled in the years since. This impressive growth was facilitated by strong pro-aquaculture governmental policies, well-organized research, education, and technology-transfer pro-grams, and increasingly open trade and access to markets. With production volumes topping 61.5 million metric tons, China alone produces more than one-half of the world's farmed aquatic foods (Food and Agriculture Organization, 2017c; Figure 3.6). All manner of production systems are used in Chinese aquacul-ture, including ponds, tanks supplied with flow-through or recir-culated water, floating net pens and rafts, submerged cages, and others, though inland ponds are the most important. More than 90 species are produced, but carps are the most important by a wide margin (Figure 3.6). These are mostly unfed, raised exten-sively in freshwater ponds, often as a secondary crop in rice pad-dies. Carps and other low-value freshwater species are still the mainstays of Chinese aquaculture, but farmers are beginning to produce more types of high-value specialty products and greater volumes of them. With lucrative export markets and continued growth of the Chinese consumer economy in mind, farmers are producing more eels, frogs, turtles, marine finfish and inverte-brates, and other high-value species. China consumes most of the fish and shellfish it produces, but does export a small fraction to the United States, Japan, and other countries.[7] Since the 1990s,

---

[7] The scale of Chinese aquaculture is such that even small percentages of production slated for export represent a significant amount of trade: for example, in 2003, Chinese exports of major aquaculture products totaled less than 2% of total aquaculture production by volume, but amounted to about 644,000 metric tons. For comparison, the meager amount that China exported was equal to more than 90% of the entire national aquaculture production of Canada that year, just under 714,000 metric tons (Food and Agriculture Organization, 2017c).

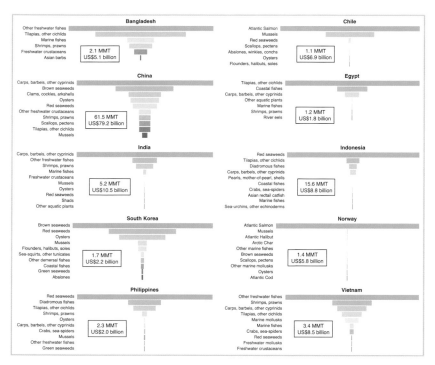

Figure 3.6: Despite the taxonomic diversity found in aquaculture, from place to place it tends to be dominated by a comparatively small number of regionally important species. The top ten species groups raised by the top ten aquaculture producing countries, shown here, illustrate the primacy of just a few finfish, shellfish, or aquatic plants in each nation's industry. These include species produced principally to satisfy domestic demand (carp in China, all finfish in India), those largely intended for export markets (salmon in Norway and Chile, shrimp in Vietnam), and others for which there is both domestic and international demand (mussels, halibut, and cod in Norway; tilapia in China) (Food and Agriculture Organization, 2017c). MMT = millions of metric tons

aquaculture has helped to increase per capita food supply and alleviate poverty, particularly in rural China, and is recognized as a significant contributor to both social and economic development (Food and Agriculture Organization, 2018a).

Although Indonesia produces only about one-quarter of the volume generated by Chinese aquaculture, with annual production volumes of approximately 15.6 million metric tons (Figure 3.6), it is the world's second largest producer of farmed aquatic organisms. Like China, Indonesia consumes most of the fish it produces,[8] but differs in that most of its production is in marine or brackish waters. Common carp were introduced to backyard ponds for subsistence aquaculture during the Dutch occupation in the mid-1800s, but Indonesians had already been raising milkfish and mullets extensively in brackish water ponds since the 17th century. All three species are important in freshwater and brackish water aquaculture today, along with Nile tilapia, Vietnamese catfishes, various penaeid shrimps, and giant river prawn. Although Indonesia is a relatively small country, the archipelago nation boasts roughly 81,000 km of coastline that has allowed mariculture to thrive in recent years, focused on groupers, pearl oysters, and katoni seaweed. A variety of culture systems are used, including freshwater (carp and other freshwater finfish) and brackish ponds (shrimps, mullets, and milkfish); floating cages suspended in rivers, reservoirs (carp and other freshwater finfish), and coastal waters (groupers); rafts and longlines (seaweed), as well as paddy field culture that integrates finfish aquaculture with rice farming. Aquaculture has become an important element of food security and economic development for Indonesia, particularly for the country's rural poor. (Food and Agriculture Organization, 2018b).

India, currently the world's third largest aquaculture producer, raises roughly 5.2 million metric tons of aquatic produce annually (Figure 3.6). Indian aquaculture is principally freshwater finfish and shellfish, and the culture of carps – particularly the Indian major carps, catla, rohu, and mrigal – catfishes, tilapias, and freshwater prawns represents approximately 95% of production (Figure 3.6). That said, there is some brackish water culture of shrimps and prawns, and a nascent mariculture sector producing small numbers of marine bivalves and finfish. With the support of publicly funded institutes and nationwide research projects to establish and disseminate aquaculture technologies and techniques throughout the country, Indian aquaculture was transformed from

---

[8] That said, much of its farmed shrimp is exported.

a subsistence activity practiced in backyard ponds to a thriving commercial enterprise. Ponds are the most commonly used culture systems, but operations span the spectrum of intensity, ranging from extensive culture involving little more than fertilizer inputs to intensive culture involving aeration and water exchanges, complete diets, health-promoting feed additives and medicines, and other inputs to support large biomasses and harvests. Polyculture (culturing multiple species together in the same system) of finfish and shellfish is common – all three Indian major carps are often cultured together or with Asian carps or giant river prawn – as is cultivation of fish in rice paddies. Linking fish production to waste management is an age-old practice in India, and treated sewage is still used today as a fertilizer in reservoirs managed essentially as aquaculture ponds for carp production. Growth of the aquaculture industry has strengthened food security in India and much of the country's farmed seafood is consumed domestically, but exports of shrimp have become an important contributor to the Indian economy (Food and Agriculture Organization, 2018c).

With approximately 3.4 million metric tons produced annually, Vietnam's aquaculture industry is the fourth largest in the world (Figure 3.6). Unlike China, Indonesia, and India where fish culture has been practiced for centuries or millennia, Vietnam does not have a long history of rearing aquatic organisms. Vietnamese aquaculture began in the 1960s with the cultivation of fish in rice paddies, ponds, lakes and rivers to contribute to the domestic food supply. Farmed fish were a particularly important source of protein during the Vietnam War. In the post-war years, aquaculture was transformed into a commercial enterprise, with an eye to the demands of lucrative export markets. In the 1980s, shrimp raised in marine or brackish water ponds became a large segment of Vietnamese aquaculture, as it remains to this day. As the 20th century came to a close the Vietnamese aquaculture industry doubled-down on export markets, developing methods to intensively rear an increasingly diverse range of finfish and shellfish desired by foreign markets, such as groupers, giant tiger prawn, lobsters, and catfishes. Diversification and promotion of high-value species continues to this day, as emerging species such as cobia, snakeheads, abalones, maculated ivory whelk, and barramundi are commercialized. Aquaculture exports to 80 different countries contribute significantly to the Vietnamese economy, and the aquaculture sector as a whole directly employs more than 670,000 people (Food and Agriculture Organization, 2017d).[9]

The Philippines aquaculture industry produces 2.3 million metric tons of farmed finfish, shellfish, and seaweeds annually, placing them fifth in the global rankings (Figure 3.6). Philippine aquaculture began with the cultivation of wild-caught milkfish fry in brackish water ponds, and milkfish remain a dominant aquaculture species, though the euryhaline species is now raised in fresh, brackish, and marine waters in both ponds and cages (Food and Agriculture Organization, n.d.). Tilapias, shrimps, carps, slipper cupped oyster, and green mussel are all important, but Philippine aquaculture is unique in that the most important farmed species are seaweeds raised for direct consumption or carrageenan extraction.[10] Staked to the bottom in shallow waters or attached to rafts or longlines in deep waters, seaweeds make up more than two-thirds of the country's aquaculture production. Like many of the aforementioned countries, aquaculture is an important source of food and income for many in the Philippines, where seaweed farming alone supports nearly 180,000 families (Food and Agriculture Organization, 2017f).

---

[9] Although 670,000 is less than 1% of the population of Vietnam (~92.7 million), aquaculture is incredibly relevant to Vietnamese employment opportunities. For comparison, aquaculture only directly supports about 10,500 jobs in the United States, a country with more than three times as many people as Vietnam (~324.1 million) (Food and Agriculture Organization, 2017e).

[10] Carrageenan is a complex carbohydrate produced in considerable quantities by many seaweeds to prevent drying out during low tides. Its gel-forming properties lend it to a variety of food technology and related applications, where it is used as a thickening, emulsifying, or clarifying agent. Carrageenan is used to keep ice cream creamy (by inhibiting large ice crystal formation during refreezing), toothpaste pasty (by keeping the ingredients from separating), beers clear (by removing cloudiness), and diet sodas satisfying (by improving texture and keeping flavors suspended throughout the solution). Carrageenan is also used in a variety of cosmetic, toiletry, and pharmaceutical products, shoe polish, and vegan/vegetarian alternatives to animal-based foods.

The Bangladeshi aquaculture industry is the sixth largest in the world, producing about 2.1 million metric tons of fish and shellfish each year (Figure 3.6). Bangladesh enjoys rich aquatic resources in the form of extensive wetlands, lakes, rivers, estuaries, and coastline as well as a diverse assemblage of roughly 795 native species of fish and shrimp. Although it is assumed that natural and manmade waterways were previously used to produce fish for subsistence or recreational purposes, there is little record of aquaculture in Bangladesh until the 1960s and 1970s when carp culture was being developed in the region. Today, pond-based polyculture of Indian major and Asian carps is common, along with pond and cage culture of tilapia and various other freshwater fish. Rearing of shrimps, prawns, and other crustaceans has expanded greatly to meet demand from domestic and foreign markets. Aquaculture has become an important contributor to the food supply in densely populated Bangladesh, particularly as water development projects have degraded aquatic habitats and negatively impacted capture fisheries in the country. Aquaculture's economic role in Bangladesh and the growing volume and value of farmed seafood exports have similarly encouraged industry expansion (Food and Agriculture Organization, 2018d).

South Korean aquaculturists raise 1.7 million metric tons of seaweed, fish, and shellfish annually, placing the country seventh in global aquaculture production (Figure 3.6). Although the country's land mass is relatively small, its territory includes 8693 km of peninsular coastline and roughly 3000 islands. Unsurprisingly, South Korean aquaculture is largely driven by cultivation of marine species, notably seaweeds, oysters, and Korean mussel which comprise the vast majority of production (Figure 3.6). These seaweeds and bivalves are primarily raised using raft, longline, or bottom culture techniques. With land in limited supply, the remaining finfish and shellfish – primarily high-value species – are raised in near- or off-shore cages or land-based tank systems supplied with flow-through or recirculated water. Although South Korea is considered an advanced economy and aquaculture's contribution to gross domestic product is relatively meager (<1%), it is nonetheless an important source of income, particularly for the rural poor. Both wild-caught and farmed seafood are essential to food security for the South Korean people, especially because of the country's limited capacity for terrestrial agriculture (Food and Agriculture Organization, 2017f).

Norway is the world's eighth largest producer of farmed aquatic organisms (Figure 3.6). Norway and South Korea are the only two countries in the top ten that are considered advanced economies, but this is where the similarities between South Korean and Norwegian aquaculture end. Whereas South Korean aquaculturists rear a rich diversity of algae, finfish, and shellfish in a variety of culture systems, Norwegian aquaculture is nearly monolithic: Atlantic salmon reared in marine net pens account for more than 99.5% of production. Aquaculture began in Norway in the mid-19th century with attempts to raise brown trout in ponds, but the modern industry was foreshadowed by successful attempts in the 1960s to rear rainbow trout and Atlantic salmon in flow-through saltwater tanks and seawater enclosures. With the development of net pens in the 1970s, aquaculture was able to take advantage of Norway's extensive coastline with its deep waters, stable temperatures, and many protected inlets and fjords. Norwegian aquaculture products are mostly exported to markets in more than 130 countries, and aquaculture is a major segment of the Norwegian economy. More than 3000 Norwegians are directly involved in aquaculture, and many others are employed by processing and other supporting industries (Food and Agriculture Organization, 2017h).

Egypt produces about 1.2 million metric tons of farmed fish annually, putting them at number nine in the world (Figure 3.6). Aquaculture predates written history in Egypt, where friezes dating back to 2500 BCE depict harvests of farmed tilapia. Although Nile tilapia is now cultivated in every corner of the world, the Egyptian native has been an industry mainstay from the very beginning. Modern aquaculture began in the 1930s with a series of governmental research and demonstration projects involving pond culture of Nile tilapia, common carp and flathead grey mullet. With continued governmental support, extensive and semi-extensive aquaculture in ponds, rice paddies, and cages grew throughout the 1970s and 1980s. In the 1980s and 1990s, aquaculture grew more intensive, land-based tank aquaculture became common, and cultivation of marine finfish began in earnest. Most of the seafood

Egypt's 65,000 aquaculturists produce is sold in domestic markets. As has been noted for most of the other countries with developing economies, aquaculture has provided Egypt's rural poor with needed sustenance and income and has reduced the cost of fish, making it a more accessible protein source for all Egyptians (Food and Agriculture Organization, 2017i).

Rounding out the top ten aquaculture producers is Chile, with 1.1 million metric tons of farmed fish and shellfish produced annually (Figure 3.6). Experimentation with bivalve culture began as early as the 1920s, but commercial aquaculture did not really begin until the 1980s. Chile is mostly coastline: the country extends 4270 km from north to south along the western coast of South America, but only averages about 177 km from east to west. Fed by the nutrient-rich waters of the Humboldt current, the Chilean coastline boasts some of the world's most productive fisheries for both native and introduced species. Nonetheless, overexploitation predictably led to declining catches, and this coupled with new economic and trade policies that favored entrepreneurial activity encouraged aquaculture development. Like Norway, Chile's aquaculture industry is dominated by Atlantic salmon cultured in near-shore net pens, though a variety of mussels, oysters, and other bivalves are also raised on rafts and longlines. Aquaculture provides a source of income to many Chileans, but feeds relatively few: aquaculture employs people mostly in rural regions, but the majority of the fish and shellfish they raise is sent to international markets, not consumed domestically (Food and Agriculture Organization, 2017j).

Aquaculture is different, almost everywhere you go. In advanced economies, the focus tends to be on intensive production of high-value species to satisfy domestic or international demand for luxury seafoods. Some developing economies have focused on species and farming methods with the fewest barriers to entry to meet basic protein demands, whereas others have looked to intensive aquaculture and lucrative international markets as a means of improving the livelihoods of their citizens. There are few constants, but wherever aquaculture is practiced, it provides food and creates economic opportunity.

Chapter 4

Why is aquaculture controversial?

Those who are reading these pages are likely well aware that aquaculture is controversial. It is unlikely that *Understanding Aquaculture* would exist if not for this controversy, or that many readers would be drawn to a title intended to address misconceptions about aquaculture. This book was written because of the debate, often misinformed, that surrounds aquaculture. Aquaculture is certainly divisive, but why is that? Terrestrial agriculture – particularly industrialized animal agriculture or "factory farming" – has its critics, and bottom trawling and some other types of commercial fishing have received their share of negative attention, but aquaculture – at the crossroads of farming and fishing – seems to have attracted a uniquely large, vocal, and diverse collection of detractors. Indeed, the arena is crowded with "a wide range of actors, including scientists, environmentalists, journalists, lawyers, local farmers, sports anglers, native communities, gourmet chefs, and so forth" (Osmundsen and Olsen, 2017: 136) and the resulting debate over aquaculture is cacophonous. To answer this chapter's eponymous question, we must consider biological, temporal, and sociopolitical factors.

Perhaps it is unsurprising that a poorly understood activity has created public concern and ire. It is perfectly natural to view what is unfamiliar, what we do not understand, with skepticism. The histories of agriculture and fishing reach far back into our past, to our days as hunter-gatherers living off the African savannas of our species' birth. Pigs were first domesticated some 15,000 years ago, followed by sheep (11,000–13,000 years ago), cattle (10,500 years ago), and chickens (8,000 years ago); cultivation of the founder crops wheat, barley, lentils, and so forth began in the midst of this ancient animal agriculture, around 11,500 years ago. Of course, fishing predates agriculture, and our evolutionary relationship with aquatic foods goes back at least 40,000 years.[1] Compared to the 250,000-year existence of *Homo sapiens* and the many millennia we have spent cultivating plants and animals on land, the 4000-year history of aquaculture is brief. Even recognizing that aquatic plants and animals have been raised in some parts of the world for thousands of years, most cultures have known aquaculture for a few centuries and then only as a relatively uncommon activity. Although farming occupies fewer and fewer of us living in the industrialized world,[2] cultivating the land is still ingrained in us, from barnyard-themed children's stories and toys, to the prized backyard gardens dotting rural and urban landscapes. Perhaps if there were a few fairytales about fish farming, aquaculture would be less hotly debated in the public square.

Part of the controversy may stem from aquaculture's unique position as the only form of animal agriculture to have significant competition in the marketplace from wild-harvested product. Today, one-half of the world's seafood is farmed – an impressive achievement, given the meager contributions of aquaculture just a few decades ago.[3] Of course, this means that the other one-half of

---

[1] The phrase "evolutionary relationship" here is apt: several researchers have suggested that incorporation of aquatic foods into the hominid diet might have been a key step in their (and our) evolution. Harvesting of mollusks, marine mammals, and fish and the associated increase in consumption of docosahexaenoic acid (DHA) and arachidonic acid (ARA) may have provided the neurological building blocks needed for the abnormally large brains *Homo sapiens* would come to possess (Crawford et al., 2000).

[2] In 1870, roughly 50% of working Americans were engaged in farming; 100 years later, that figure had dwindled to 4% (Daly, 1981). Today, U.S. employment in agriculture is less than 1.5% and continues to decline (Henderson, 2015).

[3] In the 1950s and 1960s, aquaculture contributed only about 5% to global seafood supply. Both aquaculture and capture fisheries grew steadily through the 1970s and 1980s, but capture fisheries landings leveled off in the 1990s and have remained essentially unchanged since that time. In time, aquaculture overtook fisheries as the most important source of seafood in the world. See Chapter 2 for additional details.

the seafood supply is harvested from the world's oceans and inland waters. Although hunting fowl and large game is an important source of food in some parts of the world,[4] it has largely become a recreational activity and many traditional game species are now 'ranched' if not farmed.[5] Fish is the last wild food – the only form of sustenance we still gather from the wild in meaningful amounts. Although global seafood demand continues to grow and is seemingly large enough to support thriving fisheries and fish farms, competition in the marketplace has fostered some enmity between purveyors of wild and farmed fish.

The fishing industry plays to the public's imagination in describing their work and wares. Stern-faced and with far-off eyes presumably scanning the horizon, the iconic Gorton's fisherman helms his ship and delivers Gorton's of Gloucester's frozen fish to American dinner tables. Since 1975, the company's packaging and advertising have implored consumers to "Trust the Gorton's fisherman" (Gorton's of Gloucester, n.d.). Consumers throughout North America may note a striking resemblance between the yellow slicker-clad Gorton's

---

[4] Wild meat is consumed throughout the world, but harvests are especially concerning in the tropical forests of West and Central Africa, Southeast Asia, and Amazonia. Driven by limitations to animal agriculture, the need for animal protein, and the value of wild meat in nearby markets, hunters take unsustainable numbers of small and large game, contributing to population declines and the loss of diversity in these relatively fragile ecosystems. Every week, Indonesian hunters in Sumatra harvest 25 metric tons of wild turtles and 1500 forest rats in Sulawesi. Estimates vary, but suggest that annual wild meat harvest may be as high as 23,500 metric tons in the wilds of Sarawak, Malaysia, 164,000 tons in the Brazilian Amazon, and 3.4 million tons in the Congo Basin in Central Africa (Milner-Gulland et al., 2003). Although harvesting millions of metric tons of wild game is considered unsustainable in the context of unmanaged tropical forests, the volume is nonetheless dwarfed by the 300 million metric tons of meat produced worldwide by animal agriculture annually (Food and Agriculture Organization, 2003).

[5] Bison, for example, were once hunted on the Great Plains by settlers of European descent to near extinction. Decades of wildlife management have allowed populations to recover and support recreational hunts again, but bison are also now raised throughout the western United States as a food animal in essentially the same manner as free-range cattle.

fisherman and the similarly full-bearded captain smiling from the packages of High Liner Foods' frozen seafoods (High Liner Foods, 2014). Both evoke a nostalgic view of fisherman and attach a certain wholesomeness to wild fish. "There was a romance to fishing that was inseparable from the romance of the sea, a way of life – for all its peril and terror – suffused with a freedom that the farmer and rancher would never know" (Walsh, 2011). The fishing industry has claimed and effectively defended the territories of tradition, purity, and goodness, leaving the aquaculture industry to scrabble for position on lesser, second-rate fields of play.

The fishing industry and environmental groups have become odd bedfellows in their shared opposition to aquaculture. Both have an interest in maintaining the status quo and limiting aquaculture's growth: environmental groups believe fish farming damages aquatic ecosystems and jeopardizes wild fish, and the fishing industry is in direct competition with aquaculture in the marketplace. There may be little direct cooperation between commercial fishing entities and environmental advocates, but their incentives are undoubtedly aligned against expansion of the aquaculture industry and at least some entities have been ruthlessly effective in their attacks.[6] Those representing commercial fishing and environmental protection interests are theoretically mobilized and buoyed by what economists

---

[6] Many anti-aquaculture environmental activist groups are funded by the David and Lucile Packard Foundation and the Gordon and Betty Moore Foundation. Collectively, these two nonprofits provided anti-aquaculture groups in British Columbia – such as the Coastal Alliance for Aquaculture Reform, creators of the "Farmed and Dangerous" campaign – with more than US$130 million from 2000–2011 (Krause, 2011a). The Packard Foundation has funded at least 56 organizations that promote wild salmon, attack farmed salmon, or both, to the tune of US$815 million (Krause, 2011b). The Pew Trust is one such organization that used toxicological data to undermine aquaculture and promote wild fish – see Chapter 6 for details. Although the Unites States-based Packard Foundation has a history of supporting efforts to curtail aquaculture as well as commercial fishing, their investments in British Columbian activism seem to have largely benefited Alaskan salmon fishing. Alaskan fisherman, of course, are staunchly opposed to fish farming, having consistently acted to limit growth of salmon aquaculture beginning with their successful efforts to enact legislation banning finfish aquaculture in Alaskan waters (Knapp and Rubino, 2016).

call "concentration" in public choice theory (Welch, 2015). In this context, concentration refers to the consequences – positive or negative – of an activity or policy disproportionally affecting a small segment of society. Aquaculture is in direct competition with commercial fishing and has the potential to affect the environment. Environmental groups and fishing businesses might not represent the majority of society, but the effect of aquaculture on their interests is clear – at least in the minds of their constituencies – and they are highly motivated to speak out against aquaculture. Fish farming, in contrast, typifies "diffusion", whereby many members of the public are likely to benefit from an action or policy, but the individual benefits are too small to motivate anyone to action. Many members of the public are likely to benefit from the jobs and food security that come with aquaculture, but most will only receive small, indirect benefits via stimulation of local economies, reduced seafood prices, and so on – hardly the stuff of impassioned advocacy.

Public choice theory tells us that those with the most to gain or the most to lose are the most vocal and tend to hold sway over the rest of society, even when they speak against that which would benefit the largest number of people. Naturally, the aquaculture industry has a vested interest in championing itself in the public square, but the comparatively small, diverse, and sometimes disorganized aquaculture industry has often proven no match for the disciplined efficiency of large and well-organized environmental and fishing lobbies. Few members of the public recognize they have a stake in aquaculture, so the industry has few supporters to champion it (Knapp and Rubino, 2016). During the four ill-fated attempts to pass legislation facilitating offshore aquaculture development in the USA,[7] anti-aquaculture fishing lobbyists maintained a greater presence on Capitol Hill than pro-aquaculture advocates. To this day, institutional inertia and disproportionate representation of

---

[7] The National Offshore Aquaculture Acts of 2005 (GovTrack, n.d-a) and 2007 (GovTrack, n.d-b, n.d-c) and the National Sustainable Offshore Aquaculture Acts of 2009 (GovTrack, n.d-d) and 2011 (GovTrack, n.d-e) were each introduced to the United States Congress, but died without being enacted as the 109th, 110th, 111th, and 112th Congresses gave way to their successors.

pro- and anti-aquaculture positions conspire to prevent legislative support for offshore aquaculture development in the United States. "Fishing policy … constituted a background of noise through which aquaculture issues simply could not break through. Aquaculture … was drowned in a sea of inattention" (Welch, 2015: 131). Without a doubt, the well-organized, but sometimes inaccurate campaigns against aquaculture led by environmental groups and the commercial fishing industry have made aquaculture more controversial in the public's collective mind.

In their treatise on controversy in Canadian aquaculture, Young and Matthews (2010) identified concerns regarding the environment, human health, personal rights, and rural development as four major sources of dissension. Does development of coastal waters for industrial aquaculture threaten wild fish or other parts of the marine ecosystem? Are farmed fish a healthy, wholesome food? Is it appropriate to dedicate public spaces and resources for commercial enterprise, and does aquaculture development infringe upon the rights of Aboriginal people or other citizens? Does aquaculture help rural communities? These are the questions that have driven the aquaculture debate in Canada and undoubtedly elsewhere. During a workshop on the possible development of aquaculture in the Arctic, representatives from Norway, Canada, Sweden, Iceland, and the Faroe Islands affirmed environmental issues – escapement of farmed fish and disease transmission to wild stocks for marine aquaculture, nutrient releases and eutrophication for freshwater aquaculture – as the major sources of controversy in their homelands (Karlsen et al., 2015). An assessment of the politics of aquaculture in the United States added social opposition – driven by environmental nongovernmental organizations and the media – and regulatory uncertainty to the list of sources of controversy. Media representation of aquaculture was also identified as a major contributor to controversy in Norway, but these investigators also recognized a more fundamental unease – concern over the health and safety of foods produced by the modern, industrial agricultural and food processing industries – as adding fuel to the fire (Osmundsen and Olsen, 2017). These findings were echoed in a recent synthesis of the misconceptions of aquaculture that also pointed to widespread ignorance of modern capture fisheries and aquaculture practices, concerns regarding animal welfare, perceived differences in the quality and safety of farmed versus wild fish, and trust[8] – or lack thereof – in public institutions to provide appropriate oversight as major contributors to skepticism or negative perceptions of the industry (Bacher, 2015).

Aquaculture can and does affect the environments in which it is practiced – you simply cannot produce nearly 74 million metric tons of anything without using some resources and creating some waste. Certainly, there are examples of aquaculture causing environmental damage, but scientific evidence indicates the environmental footprint of raising aquatic organisms is smaller than many activists and journalists might have you believe (Knapp and Rubino, 2016). Aquaculture is coming of age in an era of heightened environmental awareness (Figure 4.1). This is not necessarily a bad thing, but it does affect the regulatory climate and the waters of public opinion that must be navigated. Prior to seminal events such as the burning of the Cuyahoga River (1969), the Santa Barbara oil spill (1969), and the release of publications such as *A Sand County Almanac* (1949) and *Silent Spring* (1962) and subsequent passage of the National Environmental Policy Act (1970), Clear Air Act (1970), and Clean Water Act (1972), the American public was largely unaware of the environmental consequences of human activity and there was little to prevent pollution or recourse following environmental disasters.

In the decades since the environmental awakening of the 1960s and 1970s, the United States has witnessed a dramatic proliferation of environmental regulations and "green"-minded nongovernmental organizations (Figure 4.2). The same could be said for laws and organizations pertaining to animal welfare, which are becoming increasingly important in aquaculture as in other forms of animal agriculture (Huntingord, 2008). These organizations and the regulations they and the general public have supported have

---

[8] Surveys of Europeans and Americans revealed that government representatives and scientists were among the *least* trusted sources of information about seafood, with most consumers putting their faith in the hands of their family, friends, fish monger, and popular media (Pieniak et al., 2006; Hicks et al., 2008).

institutionalized an ethic of conservation across the American land-scape. Without a doubt, people and ecosystems are better protected and natural resources are better managed than in the past. That said, the prohibitions and permitting of today make it difficult for a fledg-ling industry like aquaculture to gain traction. If the poultry, swine, beef, or dairy industries were just now getting started, it is not clear whether they could operate, work out the kinks, and succeed in the current regulatory climate, particularly since they are generally less efficient in terms of edible yields and protein production and have substantially larger carbon footprints (Bacher, 2015). "'Take the time to get it right' and 'keep trying to make it better' might sound like reasonable ways to make public policies [for aquaculture]. But too much time or too many changes can stifle investment … if no projects are allowed to start, adaptive learning and improvement will have no chance to make better policies" (Knapp and Rubino, 2016: 217).

Water is essential to life, in more ways than one. Waterbodies provide us with drinking water and water to raise crops and live-stock, as well as food, energy, strategic security and defense, ship-ping routes, recreational opportunities, and more. Naturally, there is competition for water and waterways, and aquaculture must com-pete with other uses to secure water rights and farm sites. Conflicts over the right to public resources are particularly challenging for marine aquaculture. As critics are quick to point out, aquaculture can interfere with active or passive uses of open waterbodies: recrea-tional and commercial vessels must navigate around floating cages or rafts, waterfront property owners do not want their view or soli-tude spoiled by net pens and increased tender boat traffic, and so on. Whether the resource in question is a fishery, forest, or fossil fuel deposit, exploiting the commons for private gain has long been a source of controversy. Accessing public waters for fish farming is no different, though the solutions applied in the fishing, timber, and energy industries have yet to gain purchase in all parts of the world (Joyce and Satterfield, 2010; Knapp and Rubino, 2016; Young and Matthews, 2010).

The role of the media is to report the news so that the public is aware of current events and can make informed decisions about

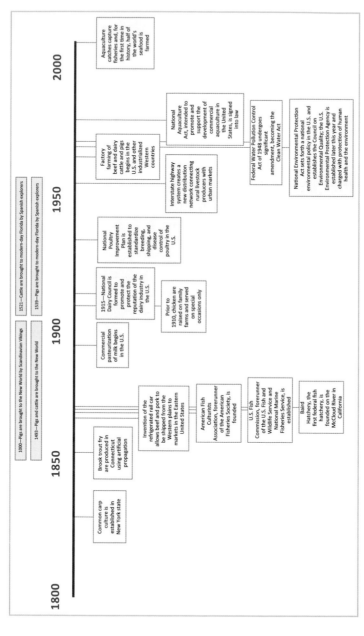

Figure 4.1: The transition of animal agriculture in the United States from a subsistence activity to big business has been punctuated by a number of scientific, logistical, and political milestones. Most of the major shifts in terrestrial animal agriculture occurred prior to the 'environmental awakening' of the 1960s and 1970s, whereas most major steps in the development of aquaculture in the United States have occurred in the decades since adoption of major environmental legislation.

Statute	Enacted	Organization	Founded
Federal Water Pollution Control Act/Clean Water Act	1948 (significant amendments in 1972)	Sierra Club	1892
Clean Air Act	1963 (significant amendments in 1970)	National Audubon Society	1905
National Environmental Policy Act	1970	Izaak Walton League	1922
Marine Protection, Research, and Sanctuaries Act (Ocean Dumping Act)	1972 (significant amendments in 1988)	National Wildlife Federation	1936
Endangered Species Act	1973	The Nature Conservancy	1951
Safe Drinking Water Act	1974	World Wildlife Fund	1961
Resource Conservation and Recovery Act	1976	Environmental Defense Fund	1967
Toxic Substances Control Act	1976	Natural Resource Defense Council	1970
Comprehensive Environmental Response, Compensation, and Liability Act (Superfund)	1980	Greenpeace	1971
Emergency Planning and Community Right-to-Know Act	1986	Ocean Conservancy	1972
Federal Insecticide, Fungicide, and Rodenticide Act	1996	Conservation International	1987

Figure 4.2: Many major environmental statutes in the United States were adopted or significantly strengthened in the 1970s or later. Similarly, many environmentally or conservation-oriented nongovernmental organizations were founded during this period. Both reflect a changing awareness within the American public and an increasing emphasis on environmental protection. Greater environmental awareness has brought agriculture and other industries into better alignment with the principles of environmental conservation and stewardship. Terrestrial animal agriculture was already well-established, but U.S. aquaculture has had to undergo its 'growing pains' in an era of greater scrutiny of environmental impacts.

social policy and their day-to-day life. Media outlets fulfill this obligation, more or less, but it is widely understood that the media also shapes public opinion directly by choosing – intentionally or otherwise – what information to relay and how to report it. Issues that are given regular coverage gather weight and importance in the public's consciousness, whereas those that are infrequently reported or ignored by the media are rarely considered, if at all. Framing – the way in which a story is told, including which facets are highlighted, what words are chosen, and what imagery and tone are used – can dramatically influence the way information is perceived. Transparent attempts to influence public opinion through framing choices are often widely criticized, but even apparent neutrality can also have an insidious effect on the public's understanding of an issue. The "we report, you decide"-style approach to journalism, whereby all perspectives are given equal consideration regardless of their connection to reality, may present itself in the guise of objectivity, but reporting opinions as facts or giving credibility and airtime to those with a complicated relationship with the truth[9] does as great a disservice to the public as overt attempts to sway public opinion through framing or selective reporting. Storylines – the condensed narratives produced by those in the media – are used in public debate as a sort of communicative shorthand. "Through storylines actors are positioned as victims, problem solvers, perpetrators, top scientists or scaremongers" (Osmundsen and Olsen, 2017: 137) and those whose storylines dominate the public debate emerge victorious from the arena. The storylines written about aquaculture – often influenced by the more colorful, emotional appeals of those lacking credentials or scientific evidence to support their positions and the incentives that drive media outlets to focus reporting on the topics that drive circulation and sales – have become entrenched and resilient to facts and new information.[10] Analysis of media reporting on aquaculture has revealed that the media itself has played a significant role in influencing how aquaculture is perceived internationally (Bocking, 2010). Media coverage has disproportionately reported the environmental and human health risks of aquaculture and farmed seafood, emphasizing the most severe outcomes, while rarely covering the

---

[9] Media coverage has often given credence to false claims about aquaculture. For example, the popular news program, *60 Minutes*, invited discredited activist Alexandra Morton, to participate in their report "Salmon in the Sea" (Gupta, 2014).

[10] The 2016 U.S. presidential election provides many excellent examples of the influence of framing and other ways in which the media can influence public opinion (Patterson, 2016). The Trump campaign slogan, "Make America Great Again" provides an especially instructive example of the impermeability of established storylines to facts: "The United States is in steep decline and we must make America great again" was a compelling and ultimately effective storyline, despite nearly all objective measures of the American economy, security, education, and so forth indicating the country had actually been on an upward trajectory for some years (Fieldstadt, 2016).

positive attributes of the industry and its products (Amberg and Hall, 2008; Schlag, 2010). Intensely negative reporting has created a number of storylines about aquaculture – it is environmentally damaging, uses more wild fish as feed than it yields, produces inferior products that are tainted with dangerous chemicals, poses a threat to wild fisheries, and so on – that are proving quite durable despite being almost completely untrue. Coverage appears to be growing incrementally more positive over time, but media representation of aquaculture and public opinion remains – at best – neutral or only slightly negative throughout much of the world, particularly among industrialized nations (Froehlich et al., 2017).

Aquaculture is widely misunderstood by those who would benefit from it, and regularly maligned by those who see it as a challenge to their own vested interests. It is a form of agriculture attempting to advance in a time when the public are increasingly removed from agriculture and invested in the green ethos of environmental protection. It is yet another industry that asks to extract value from the commons, but to what end? Who will gain, who will lose, and how much? Aquaculture is controversial, in part, because it encompasses a number of topics that are themselves the subject of considerable public debate: food safety and security, legitimacy of industrialized agriculture, animal welfare, use of public spaces and exploitation of natural resources, environmental sustainability versus economic development, and so on (Schlag, 2010). A cacophony of voices speak of aquaculture, but little is heard. In understanding what is written in these pages, perhaps the storylines can be disrupted, facts can find room in the public discourse, and the controversies can be quelled.

PART II

Health and safety issues

Chapter 5

Nutritional value of farmed versus wild fish

Every man reacts differently to a thing: his fleeting sensations cannot be expressed in any known symbols and there is no scale for determining whether a cod, a sole, or a turbot is better than a salmon trout, a fine fat pike, or even a six or seven-pound tench. (Jean Anthelme Brillat-Savarin, 18–19th century CE French politician, epicure, and gastronomic essayist.)

Fish feature prominently in folklore and mythology. Indigenous people native to the Haida Gwaii archipelago of northern British Columbia tell the story of the salmon boy, an ungrateful youth who drowns in the river and is taken by the salmon, but returns as a fish the following year to teach to his people lessons of respect, gratitude, and stewardship. According to an Indian myth with similarities to the Biblical story of Noah's ark, a small fish warns a man named Manu of a great flood and tells him to build a ship. With Manu's protection, the fish grows in size and power and ultimately guides the grateful man through rising waters to the safety of a mountaintop. As a boy, Fionn mac Cumhaill, a prominent figure in Irish and Scottish mythology, becomes all-knowing thanks to Fintan, the salmon of knowledge, after inadvertently consuming the drippings as he cooked the fish for his master. Fintan himself is said to have gained all the world's knowledge after consuming nine hazelnuts that fell from nine trees surrounding the otherworldly well in which he dwelled. The beauty and symbolism of these and other mythological fish tales have resonated across many generations, but the story of Fionn and the salmon of knowledge may have a more literal meaning we are only now beginning to fully understand and appreciate. Countless mothers have urged their children, "Eat your fish – it's brain food." While these maternal scoldings lack the poetry of Fionn and his fabled fish, they convey the same message: that seafood and intelligence are linked.

Seafood really is brain food. As it turns out, seafood and "omega-3" fatty acids have also been associated with improved cardiovascular function, optic and neurological health, fertility and early childhood development, as well as improved general health and well-being outcomes (Riediger et al., 2009). Fish and shellfish are good sources of lean protein and a number of vitamins and minerals,[1] but it is actually the fish's fat that puts it the top of nutritionists' recommendations for healthier eating. Specifically, long-chain polyunsaturated fatty acids – known to scientists as LC-PUFAs[2] – are what make seafood special. LC-PUFAs include the now-famous omega-3 fatty acids, EPA (eicosapentaenoic acid) and DHA (docosahexaenoic acid), "omega-6" fatty acids like ARA (arachidonic acid) and other fatty acids that are found primarily in seafoods.[3] These coiled, springy little molecules seem to be involved in just about every aspect of human physiology – including neurological function and brain

---

[1] Most seafoods are good sources of B-vitamins (important for energy balance and metabolism), vitamin A (vision and skin health), vitamin D (bone development and strength), iodine (essential for thyroid function), selenium (involved in preventing free radical damage), zinc (supports the immune system), and iron (supports production and function of red blood cells).

[2] Pronounced "ell-sea poof-ahs."

[3] ARA, abbreviated also as 20:4n-6, is an omega-6 polyunsaturate with four double bonds. EPA, or 20:5n-3, is an omega-3 polyunsaturated fatty acid with five double bonds; DHA is an omega-3 polyunsaturate with six doubly bonded carbon atoms. EPA and DHA take their trivial names from the Greek for 20 (*eicosa*) and 5 (*penta*), and 22 (*docosa*) and 6 (*hexa*). Despite the structural similarity between these long-chain polyunsaturated fatty acids, ARA does not adhere to the same nomenclatural convention. Rather, ARA shares a namesake with arachidic acid, a C_{20} saturated fatty acid found in peanuts and other legumes. Though ARA is not found in any plant oils, it is nonetheless named for "arachis", Latin for peanut. For additional details regarding the structure and nomenclature of fatty acids, see Figure 5.1.

power – and are what make 'good fat' good. To fully understand the importance of LC-PUFAs in human nutrition and appreciate the differences between wild and farmed seafood as sources of these critical nutrients, one must first get to know the fatty acids themselves.

Fatty acids are chains of carbon atoms surrounded – to a greater or lesser extent – by hydrogen atoms and terminated by a carboxyl group, an acidic association of two oxygen atoms and another hydrogen atom (Figure 5.1). The carbon-rich skeletons of fatty acids are rich sources of energy for cellular metabolism; this coupled with their relative stability makes fatty acids a compact means of storing chemical energy for later use. Combined three at a time with glycerol, fatty acids form the bulk of triglyceride molecules, which we recognize as the marbling in a choice cut of meat, the slippery feel of vegetable oil, and the paunch and softening of middle-aged spread. Fatty acids come in a number of shapes and sizes, depending on the number of carbon atoms they contain, as well as the number and position of any double bonds between carbons. Variation in form and physical properties make fatty acids useful building blocks for membranes and other subcellular structures where varying degrees of strength, flexibility, and permeability are required. Bonded two at a time to phosphorus-based "heads", fatty acids form the dual backbones of phospholipids. Arranged in bilayers, these sturdy molecules form the membranes that define the boundaries of all living cells. Though most of this chapter will focus narrowly on lipids[4] and their constituent fatty acids as components of food, the most elemental business of life, as we know it, could not occur without these compounds.

In addition to their roles as energy depots and subcellular building blocks, some fatty acids have additional physiological functions.

---

[4] Fats and oils are the most common examples, but lipids are a diverse group of organic molecules, defined principally by their hydrophobicity or "fear of water". Anyone who has tried to combine balsamic vinegar (a water-based solution) and olive oil (a lipid) to prepare a light salad dressing has observed hydrophobicity in action. As noted above, fatty acids are the building blocks of triglycerides and phospholipids, but related biologically important molecules such as sphingolipids, sterols, saccharolipids, and others with equally alien-sounding names also have fatty acids at their core.

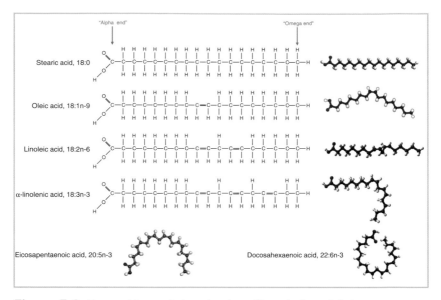

Figure 5.1: Fatty acids are chains of carbon (C symbols and dark gray spheres) and hydrogen atoms (H symbols and white spheres), terminated by a carboxyl acid group containing oxygen (O symbols and red spheres) and additional hydrogen. Fatty acids are defined and categorized according to the number of carbon atoms they contain and the number of double bonds between them. Categorizations can be further subdivided based on the position of any double bonds relative to the beginning ("alpha") or end ("omega") of the hydrocarbon chain. Stearic acid, abbreviated as 18:0, is a C_{18} saturated fatty acid, containing no doubly bonded carbon atoms. Oleic acid, is a C_{18} monounsaturated fatty acid, possessing a single double bond between the ninth and tenth carbons from the end of the hydrocarbon chain. This "omega-9" (shorthand for "omega minus 9") monounsaturate is typically abbreviated as 18:1n-9, with the numerals indicating the number of carbon atoms, number of double bonds, and position of the first (or, in this case, only) double bond relative to the omega end. Linoleic acid (LOA or 18:2n-6), is a C_{18} "omega-6" polyunsaturated fatty acid, containing two double bonds, the first of which is located between the sixth and seventh carbons from the omega end. Alpha-linolenic acid (ALA or 18:3n-3) is also a C_{18} polyunsaturate, but is an "omega 3" fatty acid with three doubly bonded carbon atoms, the first of which is located between the third and fourth carbon atoms. Eicosapentaenoic acid (EPA or 20:5n-3) and docosahexaenoic acid (DHA or 22:6n-3) are examples of omega-3 LC-PUFAs. Long-chain polyunsaturates contain 20 or more carbon atoms and at least three double bonds in their lengthy hydrocarbon chains. Although these are technically just another type of polyunsaturate, LC-PUFAs are considered to be in a class of their own because of their unique functional properties and physiological roles in living organisms.

Certain polyunsaturated fatty acids are the precursors to a variety of compounds that control myriad processes in living organisms. Eicosanoids, for example, are a family of hormone-like molecules involved in allergies and other immune responses, pregnancy and childbirth, regulating blood pressure and controlling blood flow, and various other physiological processes (Christie, 2011). These potent substances are derived from C_{20} polyunsaturated fatty acids such as EPA and ARA. Similarly, C_{22} polyunsaturates like DHA can give rise to docosanoids, a series of signaling molecules with wide-ranging, but somewhat less well-understood physiological roles (Christie, 2014). DHA is also recognized as having unique structural roles in biological membranes, particularly those of the eye and nervous system (Stillwell and Wassall, 2003). The various derivatives of LC-PUFAs are critical for proper physiological functioning in humans and other vertebrate organisms. How humans and other animals come to have the necessary LC-PUFA precursors to eicosanoids, docosanoids, and the like will eventually bring us back to the question of seafood and its nutritional value. But first, a bit more biochemistry is needed.

ARA is chemically related to linoleic acid (LOA or 18:2n-6) and, with the right enzymes, the shorter chain, doubly unsaturated molecule can be transformed into the longer chained molecule with two additional double bonds. EPA and DHA are similarly related to and derived from alpha-linolenic acid (ALA or 18:3n-3). The process begins with the insertion of a second double bond to a common monounsaturated fatty acid, oleic acid or 18:1n-9 (Figure 5.2). Desaturase enzymes create double bonds between carbon atoms.[5]

---

[5] Carbon atoms are fastidious. They like to form four bonds with other atoms – no more, no less. Sometimes, they form four single bonds with four other atoms. In other cases, they double up with one or more atoms, satisfying their organizational need for four bonds with fewer than four partners. Saturated fatty acids are known as such because each of the carbon atoms in the chain is surrounded, or "saturated" with as many hydrogen atoms as possible. Except for the carbon nearest the carboxyl or alpha end of the molecule, the carbons in the chain are bonded to neighboring carbons and make the rest of their bonds with hydrogens. For a double bond to be formed between two carbon atoms, each must sacrifice one of their hydrogen partners. The doubly bonded carbon

Delta-12 desaturase, for example, introduces a double bond between the sixth and seventh carbons from the omega end,[6] transforming 18:1n-9 into the first omega-6 polyunsaturated fatty acid, LOA. Delta-15 works similarly, but inserts a double bond between the third and fourth carbons, creating the omega-3 polyunsaturated fatty acid, ALA. These two biosynthetic transformations are not known to occur in any vertebrate animal, including humans, but are common in the biology of terrestrial plants. From these two precursors, the omega-6 and omega-3 LC-PUFAs are formed. In a series of alternating enzymatic steps, elongases add length, two carbon atoms at a time, and additional desaturases insert double bonds along the length of the hydrocarbon chain. ARA and EPA are synthesized from their respective precursors in an identical fashion, even using the same enzymes to make the necessary transformations. DHA is synthesized, in turn, from EPA according to a similar, but slightly more idiosyncratic process.[7]

Humans are not able to transform oleic acid into LOA or ALA, but if we consume these C_{18} polyunsaturates, we can complete the rest of the steps shown in Figure 5.2 to produce our own LC-PUFAs. We are not alone in this ability: if provided with dietary sources of ALA and LOA, many animals are able to produce ARA, EPA, DHA and satisfy physiological demands for these LC-PUFAs and their various bioactive derivatives. ALA and LOA are abundant

---

atoms are no longer surrounded with the maximum number of hydrogen atoms, meaning the hydrocarbon chain is now considered "unsaturated."

[6] Chemists struggle to agree on a single means of counting carbon atoms. Although 18:2n-6 is considered an omega-6 polyunsaturate because of the double bond between the sixth and seventh carbons, counting from the omega end of the molecule, the enzyme responsible for creating this double bond is known as delta-12 desaturase. Why? Those naming the desaturase enzymes counted from the opposite end of the hydrocarbon chain: whereas the double bond exists between the sixth and seventh carbons when counting from the omega end, these same atoms are the 12th and 13th when one counts from the opposite, alpha end of the molecule. Other desaturase enzymes suffer from the same identity crisis.

[7] Interestingly, the inelegant synthesis of DHA, involving an extra elongation step that necessitates a chain shortening step at the end, is considered minor evidence in support of evolutionary theory. If the process of DHA synthesis were "designed" by a god or consciousness, surely this inefficiency would have been avoided.

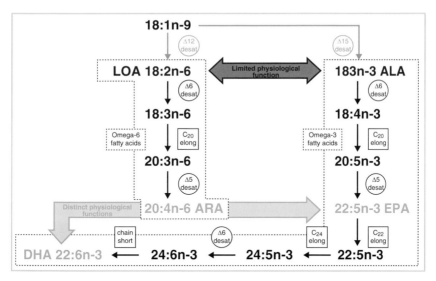

Figure 5.2: The "omega-3" and "omega-6" polyunsaturated fatty acids are chemically related and share similar biosynthetic pathways. Desaturase enzymes, which create double bonds between carbon atoms, are indicated by symbolic abbreviations in circles. Elongase enzymes (that increase hydrocarbon chain length) and chain shortening enzymes (that have the opposite effect) are shown in abbreviated form in squares. The biosynthetic steps highlighted in blue are not known to occur in any vertebrate animal; the other steps do occur, to a greater or lesser extent, depending on the species and its diet. The so-called essential fatty acids – linoleic acid (LOA or 18:2n-6) and alpha-linolenic acid (ALA or 18:3n-3) – are highlighted in purple, whereas their physiologically active derivatives – arachidonic acid (ARA or 20:4n-6), eicosapentaenoic acid (EPA or 20:5n-3), and docosahexaenoic acid (DHA or 22:6n-3) – are highlighted in green.

throughout aquatic and terrestrial food webs, so finding sources of the "raw materials" for LC-PUFA synthesis is a simple matter for these organisms. Of course, these biosynthetic machinations are not without cost, and so if these animals consume ARA, EPA, and DHA directly, there is considerable bioenergetic savings and physiological advantage to be had. Some organisms, such as felines small and large (Bauer, 2006), predatory fish like cobia (Trushenski et al., 2012) and yellowtail (Rombenso et al., 2016), and other obligate carnivores (Staton et al., 1990) at or near the top of their respective food chains have taken this strategy to the extreme: having adapted to a diet consistently rich in LC-PUFAs (found in the tissues of prey that synthesized them from C_{18} precursors) that precluded the need to expend energy to synthesize them, these animals have effectively lost the ability to transform LOA and ALA into ARA, EPA, or DHA. For these animals, LC-PUFAs themselves are considered essential dietary nutrients that must be consumed directly to satisfy physiological demand for these molecules.

Humans, with our varied diets, have retained the ability to produce our own ARA, EPA, and DHA from the LOA and ALA we consume. Although our cells must expend a bit of chemical energy to do so, this is of relatively little concern in the context of modern diets, rich in carbohydrates and other energetic fuels. Rather, our difficulty is largely related to an imbalance in the amount of omega-6 versus omega-3 precursor fatty acids. Although the process of synthesizing omega-3 LC-PUFAs involves many of the same enzymes needed to produce omega-6 LC-PUFAs, the processes are not interchangeable: it is not possible to synthesize an omega-3 LC-PUFA from LOA, nor it is possible to synthesize an omega-6 fatty acid from ALA. Only omega-3 precursors can give rise to omega-3 LC-PUFAs, and only omega-6 precursors can give rise to omega-6 LC-PUFAs. Consequently, LOA versus ALA intake determines the relative amounts of ARA versus EPA and DHA that can be synthesized in the human body. Throughout the world, the human diet has become increasingly skewed in favor of omega-6 fatty acids. Lacking the necessary omega-3 building block, ALA, or substantial amounts of preformed EPA and DHA, our bodies are unable to satisfy demand for omega-3 LC-PUFAs and instead produce more and more omega-6 LC-PUFAs.

In the approximately 250,000-year context of *Homo sapiens*,[8] the roughly 10–15,000-year history of agriculture is a relatively recent development. The agricultural revolution, followed by the

---

[8] Archaic *Homo sapiens* is thought to have diverged from its closest ancestors approximately 250,000 years ago, but hominid evolution obviously extends much, much further into the past. The genus *Homo* is thought to be at least 2.5 million years old, whereas the Hominina tribe, of which humans are the only living examples, diverged from chimpanzees and other great apes at least 6 million years ago.

development of increasingly sophisticated food processing techniques, has introduced new foods to the human diet and made others substantially more available than in the past. Whereas early *H. sapiens* were hunter-gatherers, subsisting on wild plants and animals, our modern, "postagricultural" diet – particularly in the West – is dominated by dairy products, cereal grains, refined sugars and vegetable oils, and alcohol (Cordain et al., 2005). This relatively rapid transition has dramatically altered human nutrition in a variety of ways, including the amount and, perhaps more importantly, the composition of the fats and oils we eat. With a few notable exceptions, the lipid fraction of most grain and oilseed products contains considerable amounts of LOA and relatively little ALA; as terrestrial plants are, by their nature, incapable of producing LC-PUFAs, they contain no ARA, EPA, or DHA (Figure 5.3). Consequently, by eating more and more vegetable oils, as well as the meat and milk of terrestrial livestock that have been fed corn, soy, and related feedstocks, the human diet has become exceedingly imbalanced in favor of omega-6 fatty acids.

A variety of chronic diseases and health complications have been linked to this omega-6/omega-3 imbalance (Cordain, et al., 2005), which is why health and nutrition experts throughout the world continue to urge us to consume more omega-3 polyunsaturated fatty acids – ALA, but especially EPA and DHA – to counter the effects of our otherwise omega-6-rich diets (Aranceta and Perez-Rodrigo, 2012). Nowhere are these recommendations more strident than in the case of cardiovascular disease in the West: approximately 25 of every 100 American deaths is caused by heart disease (U.S. Centers for Disease Control and Prevention, 2015); and it has estimated that eating salmon twice a week – as a source of EPA and DHA – could prevent seven of these deaths (Mozaffarian and Rimm, 2006).[9]

---

[9] But is seafood consumption in the United States on the rise? Not really. Americans consumed 15.5 pounds (just over 7 kg) of seafood per capita in 2015, which is up by nearly a pound from the previous year, but still down from the all-time record of 16.6 pounds (about 7.5 kg) per capita in 2004 (U.S. National Oceanic and Atmospheric Administration, 2016).

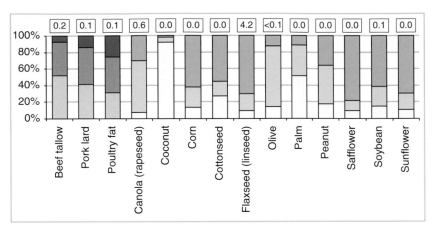

Figure 5.3: Typical composition (% of total fatty acids) of various animal fats and plant oils. Animal fats are shown in coral tones and plant oils are shown in gold tones; in both cases, saturated fatty acids are indicated by the lightest colors at the bottom of each column, polyunsaturated fatty acids are represented by the darkest colors at the top, and monounsaturated fatty acids are intermediate in both color and location. Ratios of omega-3 to omega-6 polyunsaturated fatty acids are shown in boxes at the top of each column. For many of these terrestrial lipid sources, omega-3 polyunsaturates are nonexistent or are present only in trace levels, meaning the omega-3:omega-6 ratio is effectively zero. It is important to note that in none of these cases does the limited amount of omega-3 content include EPA, DHA, or any other LC-PUFA: ALA is virtually the only omega-3 polyunsaturate found in these lipid sources. For comparison, lipids derived from fish and other seafood typically contain at least a third polyunsaturates, of which the majority is EPA, DHA, and other omega-3 LC-PUFAs.

Source: Data adapted from the National Research Council (2011).

Seafood remains one of the best sources of omega-3 fatty acids in general and, for most, it is the only significant dietary source of EPA and DHA. For the health-conscious consumer, it is reasonable to ask what type of fish is the best source of omega-3 LC-PUFAs and whether the origin of the fish – wild or farmed – matters. To answer this question – whether farmed or wild fish packs a greater nutritional punch – one must run a few simple calculations. Studies have shown that the fatty acid profile of wild fish often (but not always) contains higher percentages of LC-PUFAs than the profile of farmed fish. For example, one study reported that wild rainbow trout fillets contain 44 g of omega-3 LC-PUFAs per 100 g of fatty acids, whereas farmed trout only contain 30 g (Blanchet et al., 2005). Wild fish are the clear winner, right? Wrong. Pay attention to the units: g per 100 g of fatty acids, or in other words, percentages. Percentages get us halfway there, but one must still take total fatty acid *content* into consideration. Farmed trout, in fact farmed fish in general, tend to be fattier than their wild counterparts. The aforementioned farmed trout contained more than five times the fatty acid content of the wild fish. When you run the numbers, the farmed trout contained about 1 g of omega-3 LC-PUFAs per 100 g of fillet, whereas the wild trout only contained about one-quarter of that. Based on these figure, one would have to eat a little less than two portions (3.5 ounces or 100 g) per week of the farmed rainbow trout to meet the American recommended dietary intake level of 250 mg of EPA and DHA per day (U.S. Department of Health and Human Services/U.S. Department of Agriculture, 2015); to reach the same threshold with wild trout, one would have to consume a portion every day of the week. The same has been shown for other fish (Trushenski and DeKoster, 2017; Wright, 2016): although wild fish sometimes have higher LC-PUFA percentages, they usually have lower overall fat content, meaning fewer LC-PUFAs per portion than farmed fish.

Seafood is truly the last wild food: aquaculture is unique among animal agriculture industries in that its products compete with a near-equal volume of wild-harvested seafood in the marketplace. Consumers choose between farmed and wild seafood, but their choices are largely informed by perceptions and an inaccurate or incomplete understanding of aquaculture and capture fisheries. Many consumers are moved by their romanticized notion of wild fish. They think of lonely fishermen, married to the sea. With the knowledge of their forefathers, they outwit their well-matched but ill-fated quarry. The catch is delivered dutifully into hands that will prepare it with respect and honor. In truth, the journey of wild fish to the plate is much more likely to involve fish pumps, death by asphyxia, and mechanized processing than deft hands moving lovingly with the knowledge of generations. The mythologizing of wild-caught fish as nutritionally superior to their farmed counterparts is no more factually accurate than the folklore that began this chapter. Farmed fish are as nutritionally valuable – in some cases, more so – than wild fish, but both aquaculture and capture fisheries are needed to meet current and future demand for seafood and the omega-3 LC-PUFAs needed to counter the effects of the omega-6-dominated Western diet.

Chapter 6

Persistent organic pollutants and other contaminants in farmed fish

A little over ten years ago, Pew Charitable Trusts launched a shrewd attack against the aquaculture industry. Pew had long been critical of the aquaculture industry, but to little effect. Consumers kept on buying farmed fish and did not seem to be paying attention to their policy recommendations. But then came headlines like, "Farmed salmon more toxic than wild salmon, study finds" and "Study finds farm-raised salmon laden with cancer-causing chemicals." Suddenly, consumers' attention had turned to aquaculture and the safety of farmed fish. These and countless other headlines were the result of a US$5.5 million study, funded by Pew Charitable Trusts and subsequently published in *Science*, to assess levels of PCBs[1] and other environmental contaminants in farmed and wild fish (Pew Charitable Trusts, 2004). Research of this nature

---

[1] Originally produced for use in industrial coolants and insulating fluids, polychlorinated biphenyls (PCBs) were phased out of production in the 1970s after they were linked to neurotoxicity, endocrine disruption, and cancer, among other things. Although PCBs have not been produced in the United States for nearly 40 years, contamination remains a significant concern because of their persistence in the environment. Ironically, the chemical stability that made PCBs ideal for industrial applications is precisely what makes them so resistant to degradation and problematic in the environment. Prior to the production ban in 1979, the United States produced more than 600,000 tons of PCBs, much of which is still around. In some cases, contaminated materials have been contained to prevent exposure, but in some regions, including many of the nation's "Superfund sites", PCB contamination of the environment and biota are ongoing concerns.

is costly, but some found the price tag to be rather outsized. Further investigation of Pew's financial support revealed that the budget included US$440,000 for publicity, a highly unusual line item for a research project (Krause, 2011a).

The published article reported that farmed fish contained higher levels of some contaminants, including an eight-fold difference in PCBs, though contamination levels varied considerably in both wild and farmed fish depending on their location of origin (Hites et al., 2004). Though technically true, the relevance of a difference between 4.8 parts per billion (ppb) in wild salmon and 36.6 ppb in farmed salmon is highly questionable, given that the PCB tolerance level for seafood is 2000 ppb. The article itself was relatively careful in relating these findings,[2] but not all of the authors were so circumspect. In various interviews, David Carpenter, one of the study's co-authors said, "We hope [the study] turns people away from farmed salmon," and, "One should avoid farmed salmon like the plague. Our results indicate elevated cancer risk from one meal [of farmed salmon] or even less per month" (Krause, 2011a). The problem, of course, is that Carpenter and his co-authors found no such thing. The strongest language in the published article states that, "… this study suggests that consumption of farmed salmon may result in exposure to a variety of persistent bioaccumulative contaminants with the potential for an elevation in attendant health risks. Although the risk/benefit computation is complicated, consumption of farmed Atlantic salmon may pose risks that detract from the beneficial effects of fish consumption" (Hites et al., 2004: 227).

The nuance and speculative nature of the authors' findings were largely lost on the popular media anxious to distill the message down to simple talking points. The problem with this study is not just in the careless commentary of one co-author or oversimplification by

---

[2] Critics have pointed out a number of stylistic choices that would suggest a certain level of bias in the article. For example, farmed fish data were identified in red throughout the manuscript and wild fish were shown in green, though any colors could have been chosen to identify origin. The title itself, "Global Assessment of Organic Contaminants in Farmed Salmon", is also a bit unusual, given that the study evaluated both wild and farmed fish.

the press, but with the design of the study itself. The study was flawed in a number of ways that were likely to give wild fish the safety edge. Fish farmers, nutritionists, food safety experts, toxicologists, and others have criticized just about everything about this study, from how fish were selected and procured to how they were prepared and analyzed: farmed species were never directly compared to their wild equivalents, samples were acquired by Pew staff not the scientists involved, fish were analyzed raw and with the skin on, and very small numbers of wild fish were used. However good the intentions of those involved with this work, this just is not how an unbiased assessment of farmed and wild fish would be done. Concerns regarding the scientific merit of the research and the motives of those funding or conducting it did little to stem public fear regarding farmed fish or "The Great Salmon Panic of 2004" (Green, 2004). In the years that followed, media coverage "presented the public with a message of severe health consequences from consuming farmed salmon" (Amberg and Hall, 2008: 143). Not only is this message not supported by the findings of the studies that sparked the coverage,[3] it is demonstrably false and dangerous in that it has discouraged American consumers from eating farmed salmon and reducing the much more serious health risks associated with heart disease.[4]

The motives and methods of those involved in these comparisons between wild and farmed salmon have been questioned, but

---

[3] Amberg and Hall (2008) identified two "trigger events" that led to increased media coverage of contaminants in farmed and wild salmon: the aforementioned Hites et al. (2004) study and a dubious self-published study from the year before linking PCBs in farmed salmon to cancer and birth defects (Environmental Working Group, 2003). Interestingly, both studies were preceded by a report on levels of organic pollutants in aquafeeds and farmed salmon (Jacobs et al., 2002) that generated little interest in the United States, perhaps because of its focus on European aquaculture or insufficient resources to aggressively promote the findings in popular media.

[4] As noted in Chapter 5, roughly 25 of 100 American deaths are caused by heart disease, but many of these are preventable through dietary and lifestyle changes (U.S. Centers for Disease Control and Prevention, 2015). It has been estimated that eating salmon twice a week may reduce the number of heart disease-related deaths by more than 25% (Mozaffarian and Rimm, 2006).

this does not change the fact that seafood – wild or farmed – can contain PCBs, mercury, and other persistent bioaccumulative pollutants.[5] How does this happen? In the case of both farmed and wild fish, their food is to blame. Persistent organic pollutants, like PCBs, dioxins,[6] and dioxin-like compounds are lipophilic, meaning they bind closely to fats and oils and do not readily dissolve in water. In contaminated environments, PCBs and related compounds are typically found in the sediments. As bottom-dwelling invertebrates churn the sediments to feed, they are exposed to contamination directly and in the benthos they consume. Any PCBs they absorb become tightly bound to the lipids in their bodies and accumulate over time. In turn, dietary exposure and accumulation will continue to occur in anything consuming these prey items, thus magnifying persistent organic pollutant exposure up the food chain. With the exception of bottom-feeders, most fish are typically not exposed to these pollutants directly; rather they are exposed by consuming contaminated prey items. Biomagnification occurs in much the same way for heavy metals, except compounds containing mercury[7]

---

[5] PCBs and mercury are perhaps the most common contaminants of concern in seafood, but other persistent organic pollutants have also received considerable attention. These include a variety of dioxins and dioxin-like compounds, including intentionally produced chemicals like PBBs (polybrominated biphenyls, chemical analogs of PCBs that contain bromine instead of chlorine) and unintentionally produced byproducts of industrial processes.

[6] Dioxins have no common use or application, but were a component of Agent Orange and the other infamous "Rainbow Herbicides" used by the U.S. government in military actions during the Vietnam War. Much of Vietnam remains heavily contaminated with dioxin, and exposure has been linked with numerous health problems and deaths among Vietnamese combatants and civilians and U.S. servicemen and women.

[7] Mercury emissions can be naturally occurring (volcanic activity can release mercury to the environment), anthropogenic (resulting from human activities such as coal combustion, mining activities using mercury-based chemicals, or other chemical or industrial processes), or reemissions of previously released mercury (liberating mercury-contaminated sediments by dredging, forest fires releasing mercury previously absorbed and bound by trees). Although reemissions are thought to be the greatest contributor to contemporary mercury emissions, most of the reemitted mercury is thought to have originated from anthropogenic sources.

and other toxic metals tend to bind to proteins in living tissues, not lipids. If the ecosystem and forage base are sufficiently contaminated, fish consumption can become a significant human health risk. In 2011, there were 3710 mercury-based fish consumption advisories issued throughout the United States urging consumers to avoid or limit their intake of certain fish; 1102 fish consumption advisories were issued because of PCB contamination. Along with chlordane, dioxins, and dichlorodiphenyltrichloroethane (DDT),[8] these contaminants accounted for 94% of the 4821 advisories issued in 2011 (U.S. Environmental Protection Agency, 2011). In general, consumption of long-lived fish feeding near the top of a contaminated food chain is discouraged, whereas shorter-lived fish occupying lower trophic levels are considered safer choices (Figure 6.1).

Farmed fish are exposed to persistent bioaccumulative pollutants in the same way, except that adulterated aquafeeds take the place of contaminated prey. Various ingredients used to make aquafeeds may be contaminated with heavy metals, pesticides, or other toxic chemicals, but the most significant sources are typically fish meals and oils[9] and other animal-origin proteins and lipids (Figure 6.2).

---

[8] DDT is an organochlorine insecticide widely used to control mosquitoes (and therefore malaria) during World War II and in post-war agriculture. Although the discovery of DDT earned Paul Hermann Müller the 1948 Nobel Prize, today DDT is more commonly associated with Rachel Carson's 1962 book, *Silent Spring*, which described the dire consequences of DDT use for birds and other wildlife. Along with other galvanizing moments such as the burning of Ohio's Cuyahoga River, Carson's book is credited with the environmental awakening of the 1960s and 1970s which led to the ban of DDT, PCBs, and other ecologically damaging chemicals.

[9] Clays used in trace amounts as anticaking agents in milled feed ingredients (for example, soybean meal and the like) or to facilitate pelleting can have staggeringly high levels of contamination, depending on the type of clay and source location. Indeed, elevated levels of dioxins detected in chickens, eggs, and catfish in the United States food supply in 1997 were linked to the use of highly contaminated "ball clay" in feed ingredient milling (U.S. Food and Drug Administration, 2017b). Barring these somewhat unusual circumstances, animal fats and meals are usually the most important source of contamination in complete feeds because their inclusion rates are so much higher than the trace levels of clays incorporated in other major ingredients.

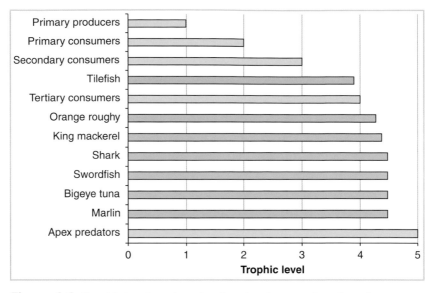

Figure 6.1: Trophic levels assigned to functional categories of producers and consumers within a marine food web (gold bars) and mean trophic levels (Froese and Pauly, 2017) for the seven seafood "choices to avoid" based on mercury contamination (U.S. Food and Drug Administration, 2017a). Note that all of the high-risk species occupy very high positions in the food web. In this example, primary producers would be mostly represented by single-celled algae whereas apex predators would be orcas and polar bears. For reference, tilapia, primary consumers with a mean trophic level of 2, are considered a "best choice."

In a comprehensive analysis of the salmonid feeds used in federal fish hatcheries in the United States,[10] various organochlorine and metallic contaminants were detected, including dioxins and dioxin-like compounds, DDT and its metabolites, and mercury. However, concentrations were below tolerance levels established for feeds for food-producing animals across the board,[11] and contaminant levels

---

[10] All of the feeds analyzed in this study were commercially available, and therefore representative of salmonid feeds used in both public and private aquaculture in the United States.

[11] In the United States, the maximum mercury tolerance in animal feeds is 2 parts per million (ppm or mg/g); the maximum concentration reported by these authors was 0.12 ppm. For PCBs, the tolerance is 200 parts per billion (ppb or ng/g); the highest residue detected in the fish feeds analyzed was 10.5 ppb

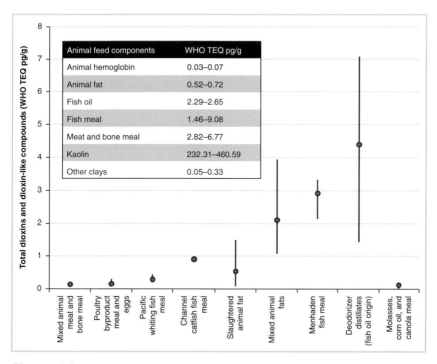

Figure 6.2: After screening a variety of United States-sourced feed ingredients researchers discovered variable levels of dioxins, furans, and dioxin-like PCBs (corrected for relative abundance and toxicity, summed and reported according to the toxic equivalency [TEQ] units adopted by the World Health Organization [WHO]) in products used to make livestock feeds, including aquafeeds (dots indicate mean contamination levels, vertical lines extend from the minimum to the maximum values observed) (National Research Council, 2003). Given that such feedstuffs and the organisms from which they are derived become contaminated through environmental exposure (or, in the case of animal-origin ingredients, contaminants in their own food supplies), it is unsurprising that levels of contamination vary among products and with geographic origin. For example, the inset table reports the range of contamination levels observed in European-origin ingredient samples (Eljarrat et al., 2002). Note contamination is reported directly as picograms per gram for some ingredients (fats and oils, mineral products), but for others (fish meal and animal-origin protein meals) values are expressed in terms of the fat these ingredients also contain, that is, picograms per gram of fat. Dioxins and related compounds are, of course, quite toxic at high exposures, but picograms are quite small: 1 picogram per gram, or 1 part per trillion (ppt), is equivalent to a moderately sized above-ground pool relative to the size of Lake Michigan.

were generally lower than previously reported for fish feeds (Maule et al., 2007). The apparent decline in contaminant levels is most likely related to the changing composition of aquafeeds, which now contain much less fish meal and oil than in the past (Tacon and Metian, 2008) and much more of the plant- and other terrestrial-origin feedstuffs commonly used to feed cattle, swine, and poultry (Figure 6.3). In fact, reducing marine-origin ingredients or limiting their application to short durations of the production cycle has been suggested as strategy for limiting aquaculture's contributions to bioaccumulative pollutants in the food supply (Berntssen et al., 2005; Crouse et al., 2013). Sparing fish meal and fish oil may provide a rare win–win – the chance to reduce contamination and save on feed cost – though the effects of ingredient substitution on the omega-3 content and nutritional value of farmed fish must still be addressed.

In the intervening years since the "salmon panic", a number of comparative reports have been released describing the contaminant burdens in a range of wild and farmed seafood. In reviewing many of these studies, species and geography were deemed much more important in determining contaminant levels than farmed or wild-origin (Domingo and Bocio, 2007). The public is right to be concerned about the safety of our food supply, which we have sadly contaminated with various types of persistent bioaccumulative pollutants. These contaminants can cause serious health problems: for example, it is estimated that dioxin exposure in food – not strictly seafood, but all foods – causes more than 193,000 illness incidents annually throughout the world (World Health Organization, 2015). Consumers should pay attention to consumption advisories for fish caught locally, and at-risk consumers, such as children and pregnant or lactating women should avoid shark, swordfish, and other long-lived fish because they tend to accumulate higher levels of contaminants. That said, in all but the rarest of circumstances, levels of contamination in fish are far, far below safety thresholds

---

(Maule et al., 2007). For those unfamiliar with these units of measure, in the context of time, 1 ppm would be equivalent to less than 1 hour over the course of 100 years; 1 ppb would be equivalent to just over 3 seconds out of a century.

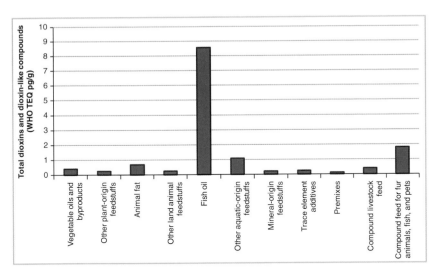

Figure 6.3: Following a number of major food supply contamination events, the European Food Safety Authority conducted an exhaustive survey for dioxin and dioxin-like compounds in common foods and animal feed ingredients collected throughout the continent (European Food Safety Authority, 2010). As has been observed in other parts of the world, aquatic-origin ingredients, particularly fish oil, are considerably more contaminated than terrestrial-origin ingredients, though the majority of these samples contained measurable levels of contamination, too. Expectedly, compound feeds for cattle, swine, poultry, and other terrestrial livestock contain little of the most contaminated ingredients and carry a lower contamination burden. On the other hand, aquafeeds and formulations for fur animals and pets routinely include fish oil, fish meal and more animal-origin ingredients in general, and thus contamination with dioxins and dioxin-like compounds in these feeds is somewhat higher. Note that all of these values are expressed as picograms per gram of product, assuming 12% moisture content for all.

and advisories. The average consumer probably is not worried about the safety of butter or brown gravy, but both are likely to contain more PCBs than farmed salmon or other fish (U.S. Food and Drug Administration, 2015a). After surveying the U.S. food supply, the U.S. Food and Drug Administration found that dioxins and dioxin-like chemicals are found in most of the foods eaten by Americans, but that the concentrations were relatively low compared to the recommended exposure limit established by the Joint Food and Agriculture Organization/World Health Organization Expert Committee on Food Additives. This international expert panel established a "provisional monthly tolerance intake" – the amount of dioxins, furans, and dioxin-like PCBs that could be consumed every month of one's life without concern for significantly damaging one's health – of 70 pg/kg body weight per day (Food and Agriculture Organization/World Health Organization, 2003), a rather high threshold given contaminant burdens reported for most foods (Figures 6.4 and 6.5).

Based on consumption rates and typical contamination levels in the United States, approximately 98% of toxic PCB exposure comes in the form of beef (~65%), milk (~19.5%), poultry (~8%), and pork (5.5%), not seafood (2%) (Green, 2004). Seafood consumption is greater in Europe, but other foods are responsible for a significant amount of dietary exposure to dioxins and dioxin-like compounds. Fish and seafood products are the most significant contributors to dioxin and PCB exposure in adolescents and adults, but milk and dairy products are the most important exposure routes for infants and toddlers; terrestrial animal meats are significant sources of contaminant exposure in all age groups (Malisch and Kotz, 2014). Some terrestrial animal products were also more likely to contain unacceptable levels of contamination: although 11–15% of fish fillet product samples exceeded established maximum levels for total dioxins, furans, and dioxin-like PCBs (depending on the TEQ calculations used), the proportion of samples in violation was greater for fat pigs (~16–22%) and mixed animal fats (~12–14%) (European Food Safety Authority, 2010). The bottom line is that both wild and farmed fish can contain measurable amounts of dioxins and dioxin-like contaminants, but so can beef and chicken and other foods with substantially less nutritional benefit to offer. About 25 of every 100 American deaths is caused by heart disease, the number one cause of mortality in the United States (Centers for Disease Control and Prevention, 2015a). It has been estimated that eating salmon twice a week might be able to prevent seven of these 25 deaths per 100. Of course, in doing so, regular seafood consumers might increase their exposure to dioxin and dioxin-like compounds and increase their risk of related health complications. What is the relative risk? If 100,000 people ate farmed salmon twice a week for 70 years, 24 might die from PCB-related cancers, but at

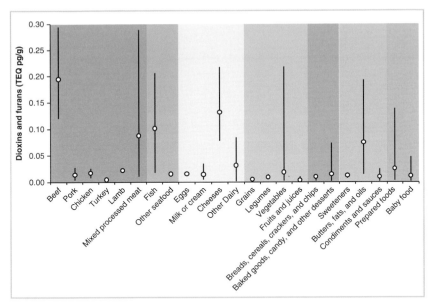

Figure 6.4: In response to concerns about dioxins and furans in the United States food supply, the U.S. Food and Drug Administration monitored the national food supply for these compounds (but, unfortunately not PCBs) throughout the late 1990s and early 2000s (U.S. Food and Drug Administration, 2015b). More than 700 different types of raw and prepared food samples were collected from throughout the country annually and measurable levels of contamination were observed in virtually every type of food (data shown were summarized from results of the last survey conducted in 2004). Lipophilic compounds like dioxins and furans tend to be sequestered in fat deposits, so it unsurprising that fattier foods such as beef, mixed processed meat (primarily luncheon meats), cheeses, and – of course – butters, fats, and oils had the highest levels of contamination. Contamination was also found in the few seafood products tested, but these were far from the most contaminated foods screened. Of course, all things are relative: the highest concentrations of dioxins and furans reported in 2004 – in oven-roasted chuck roast – were a little less than 0.3 pg/g. To exceed the Food and Agriculture Organization and World Health Organizations joint recommended monthly intake limit of 70 pg/kg body weight, an average person would have to eat approximately 14.5 kg of chuck roast every month – that is equivalent to eating nearly five portions of roast every day.

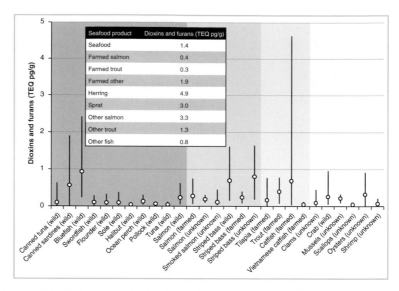

Seafood product	Dioxins and furans (TEQ pg/g)
Seafood	1.4
Farmed salmon	0.4
Farmed trout	0.3
Farmed other	1.9
Herring	4.9
Sprat	3.0
Other salmon	3.3
Other trout	1.3
Other fish	0.8

Figure 6.5: Building on the food supply surveys conducted by the U.S. Food and Drug Administration in the late 1990s and early 2000s (see Figure 6.4), a series of special studies were conducted to provide a higher resolution snapshot of dioxin and furan contamination in certain food types, including seafood (U.S. Food and Drug Administration, 2015b). A wide range of saltwater (darkest blue panel), anadromous fish (species like Atlantic salmon and striped bass that live in the sea as adults, but migrate to freshwater to spawn and rear their young; medium-dark blue panel), freshwater (medium-light blue panel), and crustaceans (lightest blue panel) were evaluated. The results revealed higher levels than observed in the full food supply survey, which largely focused on fish sticks (that is, breaded white fish) and canned tuna. Of those fish that were sampled from both wild and farmed sources, farmed fish had comparable or lower levels of contamination than their wild counterparts. The higher levels of contamination associated with farmed Channel Catfish were surprising, given the relatively short life span of these fish and the largely plant-based diets they are fed. Further investigation by the U.S. Department of Agriculture was unable to definitively prove the cause of contamination, but revealed evidence linking dioxins and furans in fillets to the same contaminant profiles in kaolin clay used as an anticaking agent (Huwe and Archer, 2013). Although the levels of contamination may seem rather troubling, it is important to note that even the most contaminated catfish fillets sampled posed a relatively meager human health risk: based on the Food and Agriculture Organization and World Health Organization recommendations, a person weighing 62 kg could consume more than nine portions of contaminated catfish per month – nearly twice a week for the rest of one's life – without exceeding the provisional tolerable monthly intake for dioxins and dioxin-like compounds (Food and Agriculture Organization/World Health Organization, 2003). The inset table reports analogous data from a survey of the European food supply (European Food Safety Authority, 2010), suggesting a similar range of contamination burdens and lower levels of dioxins and furans in farmed versus wild fish.

least 7000 would live having avoided heart disease (Mozaffarian and Rimm, 2006).[12] The risks and benefits of consuming farmed versus wild seafood are nuanced and complex: complete, accurate interpretations of the subject matter do not lend themselves to headlines. Whether intentional or otherwise, a few narrowly drawn conclusions have colored public perception of seafood and the risk of contaminant exposure to the detriment of public health.

Whereas terrestrial-origin meats are more important sources of exposure to persistent organic pollutants, seafoods are the most significant dietary contributors to exposure to heavy metals like mercury. Concern regarding mercury in seafood predates the dioxin-related food safety and public health scares of the 1990s and 2000s by at least four decades. Between 1953 and 1956, 43 residents of Minamata, Japan, died and 68 others suffered permanent disabilities as a result of methylmercury[13] poisoning.[14] Residents of

---

[12] These estimates generally do not address the influence of cooking method on contaminant exposure, which can increase or decrease contaminant exposure depending on how the food is prepared (Marques et al., 2011). Estimating dietary exposure to contaminants – through diet surveys and the like – is time-consuming and complex; assessing how much meat and seafood is baked versus fried, seared versus sautéed, and so on, is generally considered too burdensome for such studies. Where possible, such factors should be incorporated into risk assessments, but the opportunities to do so appear limited.

[13] Mercury occurs in a variety of chemical forms, all of which are potentially toxic, but some are worse than others. Pure metallic mercury and inorganic mercury compounds, such as hatters' mercury nitrate (see the next footnote), are the forms most often involved in industrial processes. Metallic and inorganic mercury are dangerous, but not as toxic as the organic form, methylmercury. Unfortunately, this more toxic form is the one that bioaccumulates in aquatic food webs after bacteria convert inorganic mercury to methylmercury.

[14] It is widely believed that the phrase, "Mad as a hatter", is an allusion to the effects of mercury exposure. For most of the 19th and 20th centuries, hat-makers prepared furs and woolen fibers with mercury nitrate, regularly exposing themselves to dangerous levels of the heavy metal. "Mad hatters" suffered the various neurological effects of mercury toxicity, including slurred speech, memory loss, impaired hearing and vision, hallucination, loss of coordination, and hand tremors. Indeed, the symptoms of mercury toxicity were so widespread among milliners that the uncontrollable tremors came to be known as "hatters' shakes" (Shackleton and Hanlan, 2004).

the small coastal city consumed great quantities of locally caught seafood and were exposed to high doses of methylmercury after decades of mercury discharges from a nearby manufacturing plant contaminated Minamata Bay and its biota. Although 2268 deaths and disablements have been attributed to "Minamata Disease", as methylmercury toxicity is now known, the National Institute for Minamata Disease has suggested that the number of those affected exceeds 18,000 (Minamata Disease Municipal Museum, 2007; Porteous, 2008).

Although the Minimata tragedy is unique in its scale and severity, mercury exposure via seafood consumption is a widespread concern. Based on the severity of potential health effects, the prevalence of exposure, and the availability of data to assess risks, the World Health Organization Chemicals and Toxins Task Force identified four heavy metal contaminants – specifically methylmercury, lead, arsenic, and cadmium – as priority substances contributing to the global burden of foodborne diseases (World Health Organization, 2015).[15] That said, mercury exposure is most concerning among a few small at-risk population segments, such as seafood-consuming women and infants living near gold-mining operations in the tropics or in indigenous Arctic populations where marine mammals are regularly eaten (Sheehan et al., 2014). Nearly all fish and shellfish contain detectable traces of methylmercury, but most seafood products do not have high contaminant burdens or pose a serious health risk (Figure 6.6) (Karimi et al., 2016) and, as a result, most populations fall into low-risk categories. By avoiding a handful of fish species and products, consumers can dramatically reduce their exposure to methylmercury. Importantly, farmed seafood may be a safer choice than wild fish and shellfish: aquaculture products generally contain less methylmercury than wild-caught products, most likely because intensive husbandry shortens the time needed for fish to reach a marketable size and can reduce the opportunities for bioaccumulation through feed formulation and ingredient selection. The same is true for other heavy metals (Heshmati et al., 2017).

---

[15] Unfortunately, the results of the Chemicals and Toxins Task Force's assessment for heavy metals is not yet available.

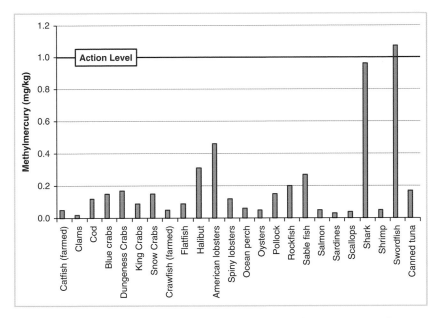

Figure 6.6: The species and products included in the figure represent the most consumed seafood in the United States by volume. Surveys conducted by the U.S. Food and Drug Administration and the U.S. National Marine Fisheries Service detected methylmercury in each of these fish and shellfish, but the vast majority of samples were far below the established action limit of 1 mg/kg or 1 part per million (ppm) (Carrington and Bolger, n.d.). The two notable exceptions, shark and swordfish, are long-lived, apex predators; feeding at the top of their respective food webs for the majority of their comparatively long lives predictably results in elevated tissue burdens of methylmercury and other bioaccumulative contaminants. A recent synthesis of various methylmercury surveys reported similar results, with only top predators including sharks, tunas, billfishes, and tilefish approaching or exceeding the 1 ppm threshold (Karimi et al., 2016).

Although there is less information available regarding methylmercury in farmed fish and shellfish, a recent review of methylmercury in the United States' seafood supply reported less contamination in aquaculture products, concluding, "… lower mercury concentrations in farmed fish compared with wild fish is broadly consistent." (Karimi et al., 2016: 1517) According to these authors, the difference in methylmercury burdens is sufficiently large and consistent to justify distinguishing the reduced exposure hazards of farmed fish from wild fish in risk assessments and the development of consumption advice.

In more recent years, Pew Charitable Trusts appears to have adopted a somewhat more pragmatic stance on aquaculture. Acknowledging that aquaculture has become the most significant contributor to global seafood supply, the Pew-funded Marine Aquaculture Task Force is now focused on "fulfilling the promise, managing the risks" of aquaculture (Marine Aquaculture Task Force, 2007). However, in a report prepared for the United States Congress, the Task Force doubled-down on claims against the safety of farmed fish, citing the Pew-funded research that instigated the salmon panic a few years before. "These studies raise serious health concerns. Their conclusions have been challenged by some in the seafood industry and others who believe benefits from eating seafood outweigh the risks. Although human health considerations are outside the scope of the Task Force's work, it must be recognized that perceptions about the health benefits and risks of eating seafood play a major role in consumer purchasing decisions" (Marine Aquaculture Task Force, 2007: 98). Indeed, "some" have challenged the findings of these studies and overblown concern regarding the risk of contaminant exposure via seafood consumption, including the Food and Agriculture Organization/World Health Organization (2011). It is telling that the Task Force considers concerns related to the human health effects of eating farmed fish outside their scope, but nonetheless questions the safety of farmed seafood and still cites the since-discredited reports. When there is little information to go on, it is easy to draw the wrong conclusion; when conclusions fly in the face of abundant data, one must consider the possibility of motivations besides those of scientific inquiry. Most seafood is safe to eat, but if generalizations are to be made, it is wild fish – not farmed fish – that pose a more significant health risk in terms of exposure to dioxins, dioxin-like contaminants, and heavy metals.

Chapter 7

Antibiotic and other drug residues in farmed fish

> Aquaculture is so festooned with antibiotics, veterinary drugs and pesticides, it can make factory farming look, well, green. (Rosenberg, 2013)

The aquaculture industry is regularly criticized for its use of medication – drugs – to treat diseases or for other, non-therapeutic purposes. This type of public concern is not unique to aquaculture – terrestrial agriculture faces many of the same criticisms leveled at aquaculture – but the arguments against the use of antibiotics, sanitizers, hormones, and other medications in aquaculture seem uniquely passionate and, in the case of the quote above, misinformed and vitriolic. Without doubt, drugs need to be used judiciously, in all forms of agriculture and human medicine. It is equally certain that treating sick fish with safe, effective medications makes good economic sense and aligns with widely held ethics related to animal welfare and the prevention of suffering. Clearly drugs are an important tool in aquaculture, but misuse can have serious consequences. In this chapter, we examine the nature and use of drugs in aquaculture, related myths and realities, and whether the industry has successfully found the middle ground between judicious use and overreliance or abuse. This exploration is largely based on the approval and use of animal drugs in the United States, but includes some comparisons with the use of drugs in other countries with sizable aquaculture industries.

First, what is a drug? You might envision pills or perhaps injections of some sort of medicine, probably an antibiotic (more on those in a bit), but "drug" carries specific regulatory meaning and the use of drugs in aquaculture encompasses a number of compounds and treatment regimens not common in human or terrestrial animal medicine.

In the United States, human and animal drugs are regulated by the Food and Drug Administration (FDA). The agency's authority is provided by the Federal, Food, Drug, and Cosmetic Act, which defines drugs as "articles intended for use in the diagnosis, cure, mitigation, treatment, or prevention of disease in man or other animals" or "articles (other than food) intended to affect the structure or any function of the body of man or other animals" (U.S. Government, 2016). In other words, if fish are exposed to _____ (insert any noun you can think of) in order for them to be _____ (insert nearly any verb or adjective you'd like), it is most likely that what is being applied meets the federal definition of a drug. Of course, this applies to what first comes to mind when thinking of drugs (that is, medicines applied to help a person or animal recover from a disease), but by this broad definition things like ice[1] and table salt[2] are also drugs. More specifically, the FDA considers ice, salt, garlic, onion, baking soda, and a host of other innocuous materials with various uses in aquaculture to be unapproved drugs of "low regulatory priority" (U.S. Food and Drug Administration, 2011); in other words, FDA is unlikely to exert its surveillance and compliance authority over these compounds, but they remain subject to federal oversight nonetheless.

The types of drugs used in aquaculture include therapeutics, spawning aids and gender manipulators, and sedatives. Therapeutics include a range of medications used to treat diseases caused by bacterial, fungal, and protozoan pathogens. Naturally occurring hormones and biochemically synthesized versions of these are used to

---

[1] Ice is commonly used in aquaculture to, obviously, reduce water temperatures and lower fish's metabolic rates prior to live hauling.

[2] Salt is also commonly used in aquaculture to help stressed fish maintain their electrolyte balance – not unlike consuming a sports drink instead of pure water after intense physical activity.

synchronize reproduction in female broodstock and, in some cases, to induce young fish to develop as female or male, regardless of their chromosomal make-up (more on this later). Sedatives are primarily used by researchers, fish health professionals, or veterinarians to keep fish calm during routine handling (such as weighing or measuring) or more invasive procedures (tissue sampling, surgery, and so forth). Although the active ingredients are sometimes different,[3] analogous products are used for similar purposes for the care of terrestrial animals, including food-producing livestock and household pets.

Whereas veterinary care of our pets is viewed favorably as a compassionate imperative, analogous medication of sick swine, cattle, or fish is often viewed with considerable skepticism regarding the safety or propriety of such treatments. Of course, we do not consume our pets, so it is understandable that people might view medications applied to animals that eventually reach our dinner tables a bit differently. That said, distrust of these practices is fueled by a fundamental ignorance of the drug approval process and, in the case of aquaculture, how approved drugs are used.

Before FDA approves a drug, they must be provided with evidence that proves the drug is effective, will not harm the intended animal, will not impact the environment, that edible products from treated animals are safe to eat, and that the drug is manufactured to be consistently potent, unadulterated, and properly packaged. These categories – effectiveness; target animal safety; environmental safety; human food safety; and chemistry, manufacturing, and controls – comprise the five "major" technical sections that along with "minor" technical sections related to drug labeling and other information must be satisfied prior to an animal drug being approved for use in the United States (U.S. Food and Drug Administration, 2015c). Each section involves rigorous standards addressing how research trials are to be conducted, how the data is gathered and analyzed,

---

[3] But not always. Florfenicol, a broad-spectrum antibiotic is marketed by Merck Animal Health under the trade names Nuflor® for FDA-approved therapeutic uses in cattle and Aquaflor® for approved uses in a number of fish species.

and submitted for consideration by FDA reviewers with specialized expertise.

This is how the drug approval process works for all animal drugs, though, in practice, more data is required for fish drugs than for other animal drugs. This is because of the large number of fish species that might be treated with a particular drug and that no single species is considered to be broadly representative of the others. In the case of cattle, there are more than 800 breeds raised throughout the world that fall into one of two broad taxonomic categories: European or taurine cattle and indicine or humped cattle. Although the European- and South Asian-origin breeds were originally considered different species, taxonomists now consider them to be subspecies of a single species, *Bos taurus* (*B. taurus taurus* and *B. taurus indicus*). Even before the two species were demoted to subspecies and merged, the diverse breeds of taurine and indicine cattle were all considered equal in the eyes of FDA reviewers. Although dairy cattle and beef cattle are handled somewhat differently with respect to drug residues in milk versus meat products and veal cattle are considered different from other beef cattle, for most intents and purposes, a cow is a cow is a cow. For example, even though indicine breeds are known to be more resistant to "foot rot" (a common bacterial infection causing lameness) than taurine breeds (Nuflor – BRD and Foot Rot, n.d.), the data required to demonstrate the effectiveness of florfenicol in support of a claim for Nuflor® does not indicate whether the cattle involved in the research trials were *B. taurus taurus* or *B. taurus indicus*, and the approved claim for Nuflor is simply for cattle, not one subspecies or another (U.S. Food and Drug Administration, 1999a).

When it comes to aquaculture drug development and FDA approval, a fish is not a fish is not a fish, and information generated in one species usually is not extrapolated to others. For instance, effectiveness data proving an antibiotic works to treat a bacterial infection in catfish is not considered proof that it would be as effective in treating the same disease in trout. One could reasonably argue that warmwater catfish and coldwater trout are sufficiently different that one should not extrapolate between these species. However, one could also argue that if catfish and trout are sufficiently similar to be susceptible to the same disease (for example, columnaris disease, a bacterial infection caused by *Flavobacterium columnare*), perhaps the other differences between these fish are not especially relevant in the context of proving drug effectiveness. Regardless of whether one considers catfish to be adequately representative of trout, most would agree that one species of catfish is a reasonable surrogate for the other species of warmwater food- and ornamental fish susceptible to enteric septicemia caused by *Edwardsiella ictaluri* (Hawke, 2015). However, because the studies conducted to prove florfenicol was a safe and effective treatment for enteric septicemia only involved channel catfish, the approval for Aquaflor® is restricted to the treatment of catfish (U.S. Food and Drug Administration, 2005).[4] Whereas FDA considers the Holstein breed to be representative of all other dairy cattle (U.S. Food and Drug Administration, 1999b), fish species are not considered to be broadly representative of others with similar life histories, rearing units, husbandry practices, and so on.

Guidance from the FDA indicates that broader, multispecies claims may be approved based on data from a limited number of species, stating, "Demonstration of effectiveness in one species from any of four broad groupings (cold freshwater, warm freshwater, cold saltwater, warm saltwater) will ordinarily be considered sufficient evidence of effectiveness against the same pathogens in all other species within that particular group" (U.S. Food and Drug Administration, 1999b: 61). However, in practice, recent drug approvals have only been granted for individual species or for warmwater, coolwater, or salmonid (not including other coldwater fish) fish groupings, and in the latter cases, usually only when data have been generated for two or more representative taxa within each temperature grouping (U.S. Food and Drug Administration, 2016). Realistically, so-called "all freshwater fish" claims are typically only possible when safety and effectiveness data are available for at least six taxa, that is, two

---

[4] The claim does not define "catfish", but it is presumed to be restricted to *Ictalurus* spp. only. Although bullheads *Ameiurus* spp. and airbreathing catfishes *Clarias* spp. are not explicitly excluded, it is unlikely that regulatory authorities would find arguments for their inclusion under the catfish claim for Aquaflor compelling.

warmwater fish, two coolwater fish, and two coldwater fish. There
has yet to be a drug approved by FDA for use in any marine fish, but
presumably the data requirements that have been used for freshwater fish would be the same for marine fish.

Because of the complexities of the drug approval process, the
volumes of data required, as well as the specialized nature of drug
approval research and its costs, the FDA has approved very few
drugs for use in United States aquaculture (Figure 7.1). If a drug is
in the development pipeline, it may be possible for farmers to use it
if FDA has granted an Investigational New Animal Drug exemption
for it (Johnson and Bosworth, 2012). There are also some limited
options for veterinarians to prescribe drugs for diseases or species
not currently on the approved uses label (U.S. Food and Drug
Administration, 2017c). For instance, if a drug were approved to
treat columnaris disease in walleye, a veterinarian could prescribe it
to treat the same infection in yellow perch. INAD exemptions and
veterinary discretion offer some flexibility in accessing a few other
fish health management tools, but it's not a free-for-all: both involve
considerable paperwork, fees, and oversight. It is also important to
recognize both approaches involve otherwise unapproved uses of
animal drugs. FDA has only granted some regulatory leeway in these
cases because of the "compassionate" need for effective treatment of
diseases in fish and other animals.

The list of approved products includes sanitizing agents containing hydrogen peroxide (35% PEROX-AID®) and chloramine-T
(Halamid® Aqua) used as therapeutic agents to treat external bacteria and parasites in fish. Although both are deemed animal drugs by
the FDA and are subject to federal regulation, neither compound
is what comes to mind when most people think of medications.
Hydrogen peroxide is a household antiseptic used to clean minor
cuts and scrapes, and chloramine is the most common disinfect-
ant used to keep U.S. drinking water safe. It so happens that, at
the right concentrations, both are highly effective in treating some
diseases in fish. Are they drugs? Yes. But are they the kind of compounds that would worry members of the public concerned about
the development of antimicrobial resistance or abuse of drugs in
animal agriculture? Unlikely.

Drug name	Active ingredient	General purpose
AQUAFLOR	Florfenicol	Therapeutic treatment of systemic bacterial infections
35% PEROX-AID	Hydrogen peroxide	Therapeutic treatment of external bacterial or fungal infections and parasitic infestations
HALAMID AQUA	Chloramine-T	Therapeutic treatment of external bacterial infections
Parasite-S, Formalin-F, Formacide-B, Paracide-F	Formalin	Therapeutic treatment of external bacterial or fungal infections and parasitic infestations
Romet-30 and Romet TC	Sulfadimethoxine and ormetoprim	Therapeutic treatment of systemic bacterial infections
Terramycin 200 for Fish	Oxytetracycline dihydrate	Therapeutic treatment of systematic bacterial infections
Tricaine-S	Tricaine methanesulfonate	Sedative for handling
Chorulon	Chorionic gonadotropin	Spawning aid
Pennox 343	Oxytetracycline hydrochloride	Marking agent

Figure 7.1: Drugs approved by the U.S. Food and Drug Administration for use in aquaculture. Note that most of these approvals are restricted to certain species, life stages (for example, eggs), or taxonomic or other fish groupings (for example, salmonids, warmwater fish). All of the approved therapeutic drug claims are further limited to the treatments of specific pathogens.

The drugs that are most likely to raise concern are antibiotics and hormones. Whether you think antibiotics are overused in livestock production or not, it is hard to make the case that they are overused in U.S. aquaculture. First, there are only three product lines of antibiotic medications that have been approved by FDA for use in fish: Terramycin® (oxytetracycline dihydrate), Aquaflor (florfenicol), and Romet® (sulfadimethoxine/ormetoprim). Notably, all of these active ingredients are commonly used throughout live-stock production, not just in aquaculture. For comparison, there are more than a dozen different types of antibiotics approved for use in chickens alone (U.S. Food and Drug Administration, n.d.). Second, the antibiotic therapeutants approved for use in aquaculture cannot be used prophylactically, that is, these medications cannot be applied unless fish are actually sick and the disease has been properly diagnosed. Third, there are no "production claims" for drugs used in U.S. aquaculture. Some antibiotics have been shown to enhance growth and feed conversion efficiency in terrestrial live-stock, and the FDA has approved claims allowing antibiotics to be used to promote growth in healthy livestock. Non-therapeutic uses of antibiotics are controversial in the public sphere, but have been relatively routine in terrestrial animal agriculture. However, there is little evidence that antibiotics have any growth-promoting effects in fish[5] and no drugs have been approved by the FDA for such purposes. In an effort to reduce risks associated with antibiotic use in animal agriculture and the development of antimicrobial resistance, on January 1, 2017, FDA placed all medically important antibiotics under veterinary oversight, including those previously available over-the-counter (U.S. Food and Drug Administration, 2017d). In addition to making all antibiotics available only via veterinary prescription or feed directive[6], FDA also worked with drug sponsors to withdraw their non-therapeutic, growth performance-related claims

---

[5] In fact, there is strong evidence that antibiotics do not promote faster or more efficient growth in aquaculture. Long-term treatment of rainbow trout, hybrid striped bass, channel catfish, or Nile tilapia with oxytetracycline has no positive effect on fish growth and may actually impair performance (Trushenski et al. 2018b). If there is no performance advantage to be had, there is no incentive to misuse antibiotics as growth promoters in aquaculture.

[6] Veterinary feed directives are not the same as veterinary prescriptions, but they are analogous in that they indicate veterinary oversight of drug use and differentiate the use of restricted versus over-the-counter medications. Whereas prescriptions are issued by veterinarians for pills, injectables, and drugs applied in water (drinking water for terrestrial animals, culture water for aquatic animals), veterinary feed directives are issued by veterinarians for drugs applied in the form of medicated feeds (U.S. Food and Drug Administration, 2015d).

for antibiotics. It is still too soon to fully know how changing access to antibiotic drugs will affect their use in terrestrial animal production in the United States or whether the tide of antimicrobial resistance can be turned. It is even less clear whether these rule changes will affect antibiotic use in aquaculture, given that no production claims were ever approved for fish, the list of approved products is extremely short, and, in the case of one drug product, veterinary oversight was already required.[7]

And what of hormones? The only hormone FDA has approved for use in U.S. aquaculture is Chorulon®, a preparation of chorionic gonadotropin used to stimulate ovulation/spermiation in broodstock and enhance spawning performance. Chorulon is also approved for reproduction-related use in cattle. The hormone treatment is typically applied via injection once, or perhaps twice at the start of the reproductive cycle to synchronize females and to increase the quality and volume of gametes produced by both male and female broodstock. Treatment does not cause maturation in fish that are not already of the appropriate age and size to undergo reproduction; rather, injected chorionic gonadotropin supplements analogous reproductive hormones produced by the maturing fish and stimulates reproductive readiness. Chorionic gonadotropin is used by commercial producers to improve the efficiency of their operations and by fisheries professionals to induce spawning in hatcheries propagating imperiled fish for restoration efforts. Hormone treatment is especially important in the latter case, where the number of broodfish is limited and collecting gametes from every individual is critical for preserving the genetic diversity remaining in the population. Although Chorulon has been approved by the FDA for use in aquaculture, it is only accessible via veterinary prescription.

Other hormones are used by the aquaculture industry to manipulate fish gender. At hatch, fish are sexually undifferentiated. Fish have adopted a number of sex determination systems throughout their lengthy and meandering evolutionary history, but for many species maleness or femaleness is defined by familiar genetic

---

[7] Since its original approval, Aquaflor has always required a veterinary feed directive for use.

conventions. In humans, most other mammals, fruit flies, gingko plants, and many, many fish, sex is governed by two chromosomes, X and Y. Individuals with two X chromosomes develop as female, and individuals with one X chromosome and one Y chromosome become male. When these fish hatch, they are already genetically male or female, of course, but have yet to develop the characteristics of being male or female. The development of ovarian or testicular tissue occurs in the following days and weeks, under the influence of the individual's chromosomal complement and genes responsible for sexual differentiation. During this short developmental window, a larval fish's genetics can be overridden by exposure to masculinizing or feminizing hormones. In other words, regardless of whether the larva possesses X or Y chromosomes,[8] it can be induced to develop as a female or male fish if it is given the right hormonal dosage at the right time. For example, larval Nile tilapia or rainbow trout will develop as male fish if their feed or culture water is treated with the right concentration of a masculinizing agent such as testosterone. Alternatively, these fish can be encouraged to develop as females if exposed to a feminizing hormone like estradiol in their feed or water (Pandian and Sheela, 1995). No masculinizing or feminizing hormones are currently FDA-approved for gender manipulation in fish, but a prescribing veterinarian can provide access to the needed drugs for restricted uses under their supervision.

Sex reversal and the plasticity of sexual phenotypes in fish is biologically fascinating, but it does beg the question: why would anyone do this? Monosex populations are advantageous in a number of ways and sex reversal facilitates the creation of all-male or all-female cohorts, either directly or indirectly. Many fish can reach sexual maturity during the course of a production cycle, but if a raceway or pond of fish only houses one sex, the possibility of reproduction is eliminated along with its associated inefficiencies.[9] Additional efficiency may be achieved by virtue of raising only male or only female

---

[8] Or W or Z chromosomes, or whatever the chromosomal equivalents for other sex determination systems.

[9] Fish usually grow more slowly after reaching sexual maturity because surplus energy is directed to ovarian or testicular development and gamete production

fish in that one sex often outperforms the other in terms of growth, conversion efficiency, dress-out, or other measures of production performance. For example, male Nile tilapia grow faster and more efficiently than their female counterparts, so food fish producers raising fish for slaughter would prefer to raise only males. Rainbow trout exhibit the opposite dimorphism, with females exhibiting the preferred traits. Hormones can be used to create monosex populations directly, such as is done with Nile tilapia: U.S. farmers regularly turn entire cohorts of larval fish male by feeding them diets medicated with 17-alpha methyltestosterone for the first four weeks of their life (U.S. Fish and Wildlife Service, 2016). Though straightforward, these treatments must be applied in accordance with strict requirements related to the handling and potential discharge of the drug product into surface water, as well as lengthy withdrawal periods preventing fish from entering the food supply before the hormone has cleared their systems.

Alternatively, hormones can be used to create the monosex populations indirectly, by creating "super brood" capable of producing only male or only female progeny (Figure 7.2). To create a broodstock that is only capable of begetting female offspring, one can simply apply a masculinizing agent to larval fish and then subsequently identify the XX fish for use as the male broodstock. Crosses between normal females and these phenotypically male, but genotypically female fish (also referred to as "neomales") will yield only female progeny. The majority of rainbow trout raised and sold in the United States are the all-female progeny of broodstocks created using these techniques. Feminizing agents can be used in a similar, though somewhat more involved process to create YY "supermales"

---

instead of somatic growth. This is particularly true among female fish, who invest much of their consumed energy in yolks and other egg-bound energy reserves that will support the early growth and development of their offspring. Sexual maturity is often accompanied by aggression and territorial behavior. At best, these interactions among competitors or between would-be mates can interfere with feeding; at worst, dominant fish may injure or kill subordinates. Finally, if attempts at reproduction are successful, there is the matter of dealing with the unwanted progeny.

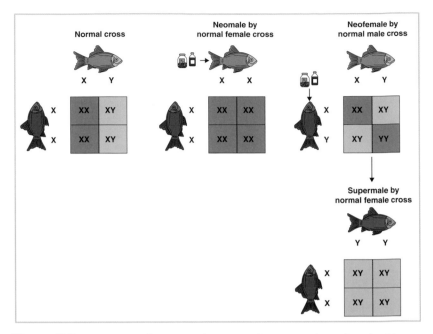

Figure 7.2: Using masculinizing or feminizing hormones, broodstock with different phenotypic and genotypic sexes can be created. When crossed with unaltered males or females, they yield progeny that are all female or all male, depending on the approach used. Punnett squares indicate the genotype of parents and progeny, whereas color is used to indicate their phenotype (blue for males, pink for females).

capable of producing only male offspring when crossed with normal females.

The indirect process of using hormones to create super brood is more involved than direct hormone treatment in the short term, but simplifies the production of monosex offspring in the long term. It is also advantageous in that it minimizes the amount of hormone needed and removes treatment from the food supply by a complete generation. Of course, as noted above, none of these hormone treatments can be applied without veterinary oversight to ensure strict adherence to requirements related to their use.

And what of drug use and oversight in other countries? Norway is the eighth largest producer of farmed seafood by volume. Norwegian aquaculture is dominated by production in marine and coastal areas, and collectively produces more salmon and trout than any other country in the world (Food and Agriculture Organization, 2016b). Norwegian aquaculture is intensive and industrialized. Given the industry's maturity and the governmental norms of Norway, it is unsurprising that aquaculture is strictly regulated. Under the authority of the Aquaculture Act 2005 and the Food Safety Act 2003 and various associated regulations, Norwegian aquaculture operations are required to operate in a manner that minimizes the risk of developing or spreading aquatic animal diseases. Operations must establish and adhere to biosecurity plans, submit to routine animal health inspections, and take immediate action involving animal health professionals when disease is suspected or in the event of mortalities (Food and Agriculture Organization, 2016c). Operators are also required to maintain records of all chemical or drug use, documenting the type of product, how much was used and for how long, and completion of any necessary withdrawal periods. The process of approving veterinary drugs and oversight of their use in Norway is guided by a regulatory framework largely analogous to that of the United States. The Act Relative to Drugs 1992 sets forth requirements for the production, testing, licensing, and distribution of veterinary drugs, including those intended for aquatic animals. Products that have been reviewed and licensed according to the standards of this regulation can then be used in aquaculture, but only under the authorization of a veterinary surgeon or, in the case of aquatic animals, a fisheries biologist, as stipulated by the Act Relative to Veterinary Surgeons and Animal Health Personnel 2001. Any treated animals are further subject to the requirements set forth in the Regulation Relative to the Control of Residues in Animal Foodstuffs, Production Animals, and Fish 2000 and the Food Safety Act 2003, which set forth restrictions on food-producing animals treated with licensed drug products, outlaw the use of prohibited substances, and the sale, processing, and consumption of animals treated with unauthorized drugs. Norwegian regulations also include provisions related to the health and safety of the environment and treated animals. By law, the Norwegian Medicines Agency and the Food Safety Authority are vested with the authority to inspect and intervene to prevent violations of the aforementioned legislation and regulations. The framework is strikingly similar to that established by the Federal Food, Drug, and Cosmetic Act in the United States, right down to the determination that salt is a chemical subject to regulation as an animal drug (Food and Agriculture Organization, 2016c).

China is the world's largest aquaculture producer by a definitive margin. Asian carps, breams, and other freshwater finfish represent more than half of the country's aquatic animal production, however, Chinese production of marine finfish, mollusks, and crustaceans is nonetheless substantial. In many cases Chinese production in these individual sectors dwarfs the entire aquaculture industries of competing countries. Control of aquatic animal diseases is addressed by the Law on Animal Diseases 1997, which deals principally with quarantine and other practices to prevent the introduction and spread of animal diseases (Food and Agriculture Organization, 2017k). Regulatory oversight of veterinary drug production, registration, and use in aquaculture is addressed generally by the Environmental Protection Law, Law on the Prevention and Control of Water Pollution, and the Marine Environment Protection Law, though the specifics are managed through a convoluted series of interrelated regulations primarily focused on toxic and dangerous chemicals. Collectively, these regulations govern the production, licensing, and use of veterinary drugs, but also stipulate that administration of drugs and drug residue tests are the responsibility of fishery authorities at or above the county level (Food and Agriculture Organization, 2017k).

India is the world's third largest aquaculture producer, but is second only to China in freshwater finfish production. Indian major carps, Asian carps, and a variety of other minor carps and freshwater fish represent 90% of production by volume, but India is also a major player in shrimp and prawn aquaculture (Food and Agriculture Organization, 2016). Despite the scale of the industry, there is no specific legislation addressing disease control or the use of drugs in Indian aquaculture. However, Statutory Order No. 722 (E) (2002) does prohibit the use of various chemicals and antibiotics in the culture of shrimp, prawn, or other fish (Food and Agriculture Organization, 2016d).

Vietnam is the fourth largest producer of aquatic organisms in the world and the third largest producer of freshwater finfish.

Although the industry is dominated by production of Vietnamese catfish, Indian major and Asian carps, tilapia, and various cyprinids, Vietnam also produces large volumes of marine finfish and crustaceans including grouper, shrimp, and lobster. The Vietnam Fisheries Law addresses prevention of disease in aquaculture through a series of provisions addressing the rearing environment, water and feed quality, and biosecurity and hygiene standards (Food and Agriculture Organization, 2016e). The law also stipulates that aquaculture operations must only use products on the list of authorized drugs in accordance with their directed use. Additionally, those involved in the use of veterinary drugs or chemicals in aquaculture must comply with the country's regulations pertaining to veterinary standards set for in the Ordinance on Veterinary Medicine. The Ordinance establishes manufacturing standards for entities involved in the production or distribution of veterinary drugs, the process for registration of new drug products, and procedures to prevent or stop the use of unauthorized drugs or drugs that have failed to uphold the quality manufacturing standards needed to stay on the authorized product list (Food and Agriculture Organization, 2016e).

Bangladesh is the sixth largest producer of aquatic animals and plants in the world, with an industry largely based on production of freshwater finfish, such as Indian major carps, Asian carps, catfish, and tilapia (Food and Agriculture Organization, 2016f). At present, there is no legislation in place to address disease control in Bangladeshi aquaculture or to regulate the use of veterinary drugs by the industry (Food and Agriculture Organization, 2016f). Though the lack of regulatory oversight is outwardly puzzling and perhaps troubling, the Food and Agriculture Organization suggests that the low intensity, extensive nature of aquaculture in this region creates little demand for veterinary drugs or other chemicals.

South Korea is the seventh largest contributor to global aquaculture production, but the industry is heavily dominated by the production of mollusks (23%) and aquatic plants (69%) (Food and Agriculture Organization, 2016g). Given this and the relative dearth of treatments and practical treatment methods for these organisms, it is perhaps not so surprising that the Food and Agriculture Organizations aquaculture legislation synopsis for South Korea is silent on the matters of animal health and veterinary medicine (Food and Agriculture Organization, 2016g). The aquaculture industry in Egypt, tenth largest in the world, is almost exclusively focused on tilapias, mullets, and catfishes (Food and Agriculture Organization, 2016h). As is the case for South Korea, there is no information available via the Food and Agriculture Organization regarding regulation of aquaculture disease control in Egypt. Although Law No. 4/1994 generally prohibits the handling of hazardous materials, including pharmaceuticals without a permit, there does not appear to be any specific regulation with respect to the use of veterinary drugs in aquaculture or livestock sectors (Food and Agriculture Organization, 2016h).

The aquaculture industry in the Philippines, the fifth largest in the world, is dominated by production of aquatic plants (66%), but also produces a substantial volume of freshwater and marine finfish (Food and Agriculture Organization, 2016i). Perhaps driven by the larger scale of the industry and greater emphasis on finfish, the Philippines does have considerable infrastructure and legislation to address aquatic animal health and the use of veterinary drugs (Food and Agriculture Organization, 2016i). The Fish Health Section of the Bureau of Fisheries and Aquaculture Resources monitors the aquaculture industry in the Philippines. The Fish Health Section engages in disease surveillance, conducts aquatic animal health inspections and issues certifications, and provides technical support to aquaculture producers in the form of diagnostic services and guidance related to veterinary drugs through designated Fish Health Officers, some of whom are also deputized as Veterinary Drug and Products Control Officers. The drugs themselves are subject to the oversight of the Bureau of Animal Industry, which oversees the production, licensing, and use of drugs used in food-producing animals, including aquatic species (Food and Agriculture Organization, 2016i).

Indonesia, second only to China in total aquaculture production, is also divided between the production of aquatic plants (70%) and animals (30%) (Food and Agriculture Organization, 2018e). Aquatic animal production is dominated by carps, tilapias, and other freshwater finfish, but Indonesia also produces significant volumes of marine/brackish finfish like milkfish and grouper, as well as various marine shrimps and prawns. Disease control is addressed primarily by Law No. 16 concerning Animal, Fish and Plant Quarantine that, along with other related legislative decrees, establishes mandatory quarantine requirements including disease treatment for live aquatic animals imported to Indonesia. Interestingly, these measures do not apply to exported animals or movement of animals within the country. However, these requirements are mandatory only for applicable quarantine diseases, as defined by the competent minister. Although the 2004 Fisheries Law forbids the use of substances that might harm human health, aquatic resources, or the environment and there is a certification system in place to ensure quality and safety in animal drug production, there are no specific provisions addressing the use of drugs in Indonesian aquaculture.

Chile is the ninth largest aquaculture producer. Like Norway, the Chilean aquaculture industry is dominated by saltwater-reared salmonids and other marine species (Food and Agriculture Organization, 2018f). The General Fisheries and Aquaculture Law and subsidiary regulations address aquatic animal health primarily through import and quarantine requirements, fish health certification procedures, and required actions when a disease is suspected or confirmed, including provisions ranging from surveillance and disinfection to depopulation and restrictions on when naïve populations can be reintroduced to the affected area. The Animal Health Protection Law, Agriculture and Livestock Law, and their subsidiary regulations address the production and use of authorized and experimental veterinary drugs, however, none of these or the aforementioned regulations specifically address the use of drugs in aquaculture.

As the preceding paragraphs illustrate, what constitutes legal use of drugs in aquaculture varies substantially from country to country. Indeed, there is great variation in whether or not these countries – the top ten aquaculture producers in the world – even have the legislative or regulatory language in place to define acceptable practices. Equally important is the question of enforcement, which is difficult to gauge: a country may have laws and regulations in place, but if there are no means to assess compliance and issue fines or other penalties for violations, there is little to deter those less conscientious aquaculture operators from misusing veterinary drugs.

Perhaps the most pressing concern among consumers is not whether farmed fish or shellfish have been exposed to an illegal substance or an otherwise improper drug application, but whether the resulting seafood bears any dangerous drug residues. Before a drug is approved for use in the United States, the FDA must be provided with a wealth of data to establish patterns of drug residue depletion and conservative withdrawal periods to prevent unsafe foods from entering the human food supply. Using highly sensitive methods to detect even the faintest traces of a drug, researchers monitor the blood, muscle, and other tissues of fish following experimental drug treatment. Using these data and pharmacokinetic (PK) models,[10] they characterize how a drug moves through a fish's body and, assuming "worst-case scenario" conditions,[11] establish conservative withdrawal periods to ensure the drug and its metabolites are fully depurated from the edible tissues.

How conservative are these withdrawal periods? Take the aforementioned florfenicol, for instance: the current U.S. tolerance for florfenicol in food is 1 ppm.[12] Following treatment, florfenicol residues drop to 0.3 ppb or less[13] within 2 weeks of treatment in a

---

[10] Studies investigating the metabolic fate of a drug in a living organism are referred to as pharmacokinetics. So-called "PK" work typically involves identifying all of various chemical derivatives an animal might produce during its metabolism of the drug in question. Researchers then measure concentrations of these metabolites and the "parent compound" in the tissues of fish, monitoring the initial increase and subsequent decrease in tissue levels in the hours, days, and weeks after treatment.

[11] For most drugs, the worst-case scenario for drug clearance involves cold water temperatures and fish with fattier tissues. Because fish are poikilothermic, their metabolism and drug excretion rate are slower at colder temperatures. Additionally, many drugs are lipophilic (literally "fat loving") and are more difficult to excrete when there is an abundance of fat in the tissues to embrace and envelop the eager molecules.

[12] Or, 1 mg of florfenicol per kg of food. For those unfamiliar with these measures, this is roughly equivalent to 2 or 3 grains of table salt per bottle of wine (including the weight of the bottle).

[13] 0.3 parts per billion, or 0.3 micrograms per kg, is more than 3300 times lower than the 1 part per million tolerance. Again, for those unfamiliar with these measures, this would be roughly equivalent to the weight of a check mark written in pencil relative to the weight of a gallon of milk.

variety of warm- and coolwater fish; in other species, tissue levels drop below detection within 96 hours. Despite these rapid depuration rates, fish treated with florfenicol must complete a 15-day withdrawal period before they can be consumed or released to the environment (Bowker and Trushenski, 2015). Compared to the rate of drug clearance, the florfenicol withdrawal period is almost absurdly lengthy, but FDA's standard is not whether fillets of treated fish have residues that are "low enough". Rather, their intent is to prevent adverse effects in a person consuming the fillets of treated fish every day of their life.

Despite lengthy withdrawal periods and legal barriers to prevent misuse of drugs in aquaculture, drug residues are occasionally detected in the seafood supply. The seafood inspection programs in Canada, the European Union, Japan, and the United States reported relatively few instances of actionable drug residue detections in domestic or imported products (Love et al., 2011). For example, FDA seafood inspections detected a total of 138 violations over a five-year period, corresponding to less than 0.1 violations per 10,000 tons of edible seafood per year. Of these violations, only 10% were associated with U.S. domestic seafood. Most of the violations were associated with seafood imported from Vietnam, China, and to a lesser extent, Indonesia. These results are corroborated by an assessment of seafoods after they made it to market: a variety of mostly farmed seafood available in the American Southwest were screened for 47 different antibiotic residues, but only five of these compounds were detected. Of those detected, each was at levels well below established tolerances set by FDA. With little evidence of antibiotic residues, much less dangerous levels thereof, these researchers concluded that there is relatively little risk of antibiotic exposure as a result of seafood consumption in the United States (Done and Halden, 2015). It is not clear whether Chinese, Vietnamese, or Indonesian fish farmers use different production methods and treatment scenarios based on whether they are targeting domestic or export markets, so the results of this survey of Asian seafoods in the United States may or may not be representative of the farmed seafood consumed in the producing countries.

In a recent article, appropriately titled "Mythbusters", Jim Bowker and I noted that, "… without having the slightest notion as to how, why, or even if it's true, the general public simply 'knows' that cultured fish are burdened with drug residues. As it happens, there are no scientifically valid, statistically defensible data to support such claims relative to elevated tissue drug residue levels, and yet the myth is pervasive" (Bowker and Trushenski, 2015: 360). Jim recently retired from his position as the Research Program Manager for the U.S. Fish and Wildlife Service's Aquatic Animal Drug Approval Partnership Program, the only dedicated aquaculture drug approval entity in the United States. He has been a part of every major drug approval effort in the United States for the past 20 years and knows all too well how aquaculture drugs and his work to make more drugs available are perceived.

> When people would ask me what I do for a living, I used to try to explain that the U. S. Food and Drug Administration considers fish a food animal, that only FDA-approved drugs can be administered to fish, and my job was to generate data to support drug approvals for fish. Over and over, the reaction I got was, 'You use drugs on fish?! What do you mean FDA has to approve fish drugs?' So I started simplifying my message to say that I was involved in fish health and doing work to try to make sick fish better. More recently, I've seen a few thought-provoking 'Ted Talks' about communicating science to non-scientists that have helped me to provide a more compelling explanation, avoid jargon, and use terms that others are more comfortable with. My 60-second elevator speech now consists of telling folks that I work in a program that helps to ensure that fish raised by hatcheries and fish farms in the U.S. are healthy and wholesome and that we work for the only program in the country focused solely on getting medicines approved for use on fish (J. Bowker, interview with J. Trushenski, 2016)

At least some of the confusion and reactions that Jim encounters are fueled by misconceptions about drug use and residues in aquaculture. "I fly fish, and many of my fly fishing friends are what I would call purists – they tend to not support aquaculture. They say things like 'I wouldn't eat a hatchery fish because it would be soft and mushy,' or, 'I know I caught a hatchery fish because it didn't fight very well.' I can only imagine what else they think about aquaculture, especially when it comes to drug treatments – I'm guessing there are a few 'alternative facts' in there. I try to turn these occasions into teachable moments, but that is not always an easy thing to do, especially when you see how often popular media, bloggers, etc., get it so wrong."

Over the course of several conversations, Jim shares "tall tales" he has heard about drug abuse in aquaculture: farmed fish are swimming in antibiotics, they are filled to the gills with dangerous drug residues, and so on. Of course, when you dig a little deeper, you regularly find that the facts tell a different story or the facts are absent altogether. "When we were researching the Mythbusters article, we found nothing that would suggest that aquaculture drug use in the U.S. is a problem. Suggestions of antimicrobial resistance turn out to be anecdotal or equivocal, or occurring elsewhere where drug use is unregulated. Detectable drug residues are reported in U.S. seafood, but it turns out these fish were raised somewhere else and shipped here for processing." That is not to say that misuse of drugs in aquaculture does not happen at all or that the consequences are not serious consequences, it just does not seem to happen very often. "I have visited dozens and dozens of public hatcheries and private farms over my career and have come to know many of those involved in commercial, tribal, state and federal agency aquaculture, and I am continually impressed with their level of commitment and how conscientious they are about adhering to the rules for legal and judicious use of drugs. Word gets around fast, so if illegal or unwarranted drug use were happening regularly in the U.S., you'd hear about it."

Jim strongly supports aquaculture, especially the domestic U.S. industry, but does he put his money – or farmed seafood – where his mouth is? Laughing, he admits that he does not eat seafood as much as he should. "I'll admit that I love a great steak and that would usually be my first choice, but I do order fish when I go out to dine." And does he worry about drug residues? "I don't think I've ever considered drug residues while reading a menu. If I really was worried about that sort of thing, I would probably shy away from land-based meats more than fish, just given the difference in the volume of drugs used in terrestrial livestock versus fish. The withdrawal periods established by the FDA are designed to ensure that the animal's meat is safe to eat not just that day, but every day for your entire life. That is a pretty high standard."

Jim is more aware of what drugs are used in aquaculture, how they are applied, and how the U.S. food supply is monitored to ensure violations are detected than virtually anyone else you are likely to encounter, even within specialized aquaculture or drug approval circles. So what would he like the rest of us to know? How would he sum up his own Ted Talk about drug use in aquaculture and the safety of farmed fish? "In the 20 years I've been involved in drug development and use in aquaculture, I haven't seen anything that would make me uncomfortable ordering farmed surf instead of turf."

Chapter 8

Genetically modified organisms in aquaculture

Brightly colored and beautiful, Glofish® are quite popular in the aquarium trade (Figure 8.1). They delight children and adults alike, but these cosmically colored specimens have a secret. They are genetically modified organisms (GMOs).

As one might expect, given their somewhat fantastic names, "galactic purple zebrafish", "moonrise pink tetras", and "electric green barbs" are not naturally occurring, but the result of a handy bit of genetic modification. Although more than 180 fish are known to be naturally biofluorescent, tiger barb, black tetra, and zebra danio are not among them (Sparks et al., 2014). In fact, these normally drab fish are pet store staples, inexpensive and hardy choices for new aquarium hobbyists. But splice in a gene from a jellyfish, coral, or anemone and a promoter that encourages expression of this gene in muscle cells, and the once ordinary fish are transformed. Newly galactic and sunbursting, these fish grow flush with the colors of fluorescent cnidarian proteins churned out by their own piscine ribosomes, just the way nature intended it … well, nearly.

Glow-in-the-dark pets might seem like a frivolous application for genetic engineering; indeed, some would question the ethical propriety of such an endeavor. Creating GMOs is no small feat, and developing a line of novelty pets was certainly not the objective of those working to create a fluorescent transgenic fish. Glofish were actually developed as part of an effort to create sentinel fish that would fluoresce when exposed to environmental contaminants (Yorktown Technologies, L.P., 2017). Indeed, there is still considerable interest

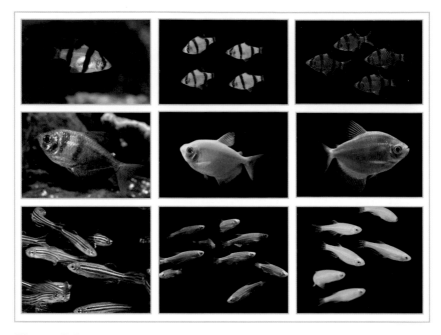

Figure 8.1: Examples of wild-type tiger barb, black tetra, zebra danio (left) and their genetically modified Glofish® counterparts (center and right). Normally drab in appearance, the bright colors of Glofish tetras, barbs, and danios are the result of inserted genes from jellyfish, corals, and other bioluminescent invertebrates.

Source: All Glofish images are courtesy of Yorktown Technologies, L.P.; all other photos are credited as follows: tiger barb © Author: Malene Thyssen / Wikimedia Commons / CC-BY-SA-3.0/GFDL; black tetra © User: Serentier / Wikimedia Commons / CC-BY-SA-3.0; zebra danio, © Author: Oregon State University / Wikimedia Commons / CC-BY-SA-2.0.

in using these GMOs in ecotoxicology (Lee et al., 2015). For the creators of the Glofish technology, the commercialization of their model organisms is just a bit of fun and commerce on the side. Glofish remain the only genetically modified animal publicly marketed in the United States, though the creation of fluorescent animals for research purposes has become relatively commonplace. To date, scientists have used genetic modification techniques to create fluorescent specimens of nematode, fruitfly, various fish, frog, mouse, rabbit,[1] and

---

[1] Unlike the other fluorescent genetically modified animals created for scientific purposes, "Alba", the transgenic rabbit was commissioned as part of an elaborate

monkey (Stewart, 2006). Of course, the controversial story of genetic engineering did not begin with Glofish or any of these other glow-in-the-dark organisms. In fact, much of the public furor over GMOs does not involve transgenic animals at all, but rather the transgenic plants that have displaced traditional corn, soybeans, and other crops throughout much of the world.

Glofish and the various other neo-fluorescent animals mentioned above are examples of transgenesis, whereby the genome of one organism is altered by the introduction of DNA from another species. There are other types of genetic modification, but when people talk of GMOs, it is typically transgenic organisms to which they refer. That said, genetic modification may refer to any direct modification of an organism's genes, including manipulation of its own genes. Under normal conditions, bacteria routinely exchange genetic material by trading plasmids, circular loops of DNA. Bacteria also naturally produce enzymes that chop up DNA molecules as a means of defending themselves against invading viruses. By harnessing these natural processes and combining them, researchers in the 1970s were able to excise genes and introduce them to recipient bacteria via modified plasmids. Techniques were refined and expanded, and commercial applications soon followed, with transgenic bacterial lines being created to produce renin, an enzyme used in cheese-making, and insulin, a life-saving hormonal treatment for diabetics (Turner, 2013). Genetic modification graduated to more complex organisms in the ensuing decades. In the 1980s, *Agrobacterium tumefaciens*, a bacterium that naturally introduces its genetic material into plants, was brought to heel and used to introduce genes of interest into the genomes of plant cells. Later, a new process was developed to create transgenic plants by physically bombarding cells with the DNA of interest. Around the same time, genetically

---

artistic effort intended to spur examination of what some might consider to be the Pandora's box of genetic modification, social and ethical responsibility in transgenesis, what it means to be human, and the nature of life itself (Kac, 2000). Unsurprisingly, the project was controversial and, as the scandal grew, the collaborating geneticist responsible for creating Alba ultimately distanced himself from the project.

modified animals were being created by literally injecting DNA into the early embryonic stages of mice and other species. Commercial applications were soon developed, most notably in the form of herbicide-resistant crops or crops that produced their own insecticides,[2] and experimental organisms with genetic predispositions that made them useful models for biomedical research. Though genetically modified (GM) research animals have become a mainstay for many research communities, global use of genetic modification is dominated by the production of transgenic soybeans, corn, and other crops, which represented 90 million acres of farmland in 2005. The majority of corn and soybeans raised in the United States is genetically modified, and derivatives of these crops find their way into an estimated 70–75% of processed foods (Turner, 2013).

This is not to say that acceptance of GM foods is widespread. Indeed, the safety and ethical propriety of GM crops is one of the most controversial issues of the past decade.[3] Vast amounts of GM

---

[2] The first GM crop to be approved in the United States, the slow-ripening, long-lasting Flavr Savr™ tomato, was a commercial flop. The Roundup Ready® soybean, followed by other Roundup Ready crops, proved substantially more successful. Developed by Monsanto, Roundup Ready crops tolerate the herbicide Roundup® (also produced by Monsanto) allowing entire fields to be sprayed with the herbicide without affecting the crop. Soon after, Monsanto developed the Bt series of transgenic crops. Armed with the genetic material from *Bacillus thuringiensis*, Bt crops are toxic to a variety of insect pests, precluding the need for insecticide application in the field. Today, Monsanto produces numerous genetically modified seed lines, including lines with multiple transgenically introduced traits (Turner, 2013). GM crops are not the first of Monsanto's many products to draw public disapproval. Monsanto is infamous for having produced industrial chemicals such as PCBs, pesticides including DDT and Agent Orange, and recombinant bovine growth hormone, all of which are now associated with a variety of insidious effects threatening human and environmental health. Given the company's controversial history and market dominance in the field of genetic engineering, it is likely that at least some of the criticism of genetic modification is rooted in negative perceptions of Monsanto, DuPont, and other biotech giants, large corporations in general, and the industrialization of agriculture.
[3] Much of the concern over the safety of consuming GM foods is based on hypothetical or equivocal information (Hilbeck et al., 2015), but this does not mean that risks are nonexistent or not worth investigating. What the debate lacks in dispassionately and rigorously evaluated data, it makes up in hyperbole:

soybean, corn, cotton, and others, are raised in the United States, Argentina, and Brazil, the top three producers of GM crops. In these countries, as well as China, Canada, Mexico, and other countries allowing these crops, GMOs are subject to a safety evaluation process and must be approved prior to commercialization. Other countries have adopted a more skeptical, restrictive approach. For example, European Union regulations stipulate that only approved GM crops may be grown and foods containing GM ingredients must be labeled accordingly; South Korea regulates GM crops in an analogous fashion. European Union member nations are permitted to adopt more rigorous restrictions: Germany, Austria, Hungary, Greece, Bulgaria, Poland, and Luxembourg have all effectively banned or "opted out" of GM food production within their borders, though many of these countries import GMOs to feed livestock or process into human foods. Countries such as Israel and the Russian Federation do not allow GM crops to be raised, but imported GMOs are significant contributors to these nations' food supplies. Certain GM crops are legal in Japan, but none are grown as a result of public safety concerns; paradoxically, Japan is one of the largest importers of GM crops (The Law Library of Congress, Global Legal Research Center, 2014).

Of course, transgenesis and other forms of genetic modification are not the only way in which organisms have been altered as a result of human intervention. To suit various purposes and desires, humans have been altering the genetic make-up of plants and animals for millennia through the process of selective breeding or artificial selection. Though the process is glacially slow in comparison with direct modification of an organism's genome, the results are no less dramatic. One has only to compare any of the 177 recognized breeds of *Canis lupus familiaris* to modern-day wolves or their shared canid ancestors to see the power of selective breeding (Figure 8.2).

---

suggesting that the safety of GM foods has not been adequately investigated, one critic said, "We are now being asked to believe that everything is OK with GM foods because we haven't seen any dead bodies yet" (page 1041; Butler and Reichhardt, 1999).

Figure 8.2: Generations of selective breeding and domestication applied to ancestral canids led to the divergence of domestic dogs and closely related wolves. Domestic dogs *Canis lupus familiaris*, such as the chihuahua (middle) and shar-pei breeds (right) are strikingly different in form and behavior from wolves, such as the gray wolf *Canis lupus lupus* (left).

Source: gray wolf, created by Gary Kramer and released into the public domain by the U.S. Fish and Wildlife Service; chihuahua, © Author: Alfredo Villa / Wikimedia Commons / CC-BY-SA-2.5, shar-pei released into public domain by User: Gothika / Wikimedia Commons.

Similar examples of transformation can be found throughout plant and animal agriculture. Generations upon generations of selective breeding have given rise to lines of broiler chickens that go from egg to plate in 50 days or less and a multitude of maizes used for feedstock, grinding into meals, or eaten off the cob. Broiler chickens are heavier and less flamboyantly colored than red junglefowl, but still bear some resemblance to their progenitors. Maize, in contrast, is nearly unrecognizable in comparison with the wild teosinte grasses that share the genus *Zea*. The result of repeated selection compounded over many generations is striking, but nothing more than a harnessing of natural selection and the evolutionary yearning toward ever-greater biological fitness. To most people's minds, selective breeding is a far cry from gene splicing.

Other examples of genome manipulation are closer to genetic modification, but still far short of the biotechnological introduction of one species' genes into the genome of another. Hybridization, the interbreeding of closely related, but distinct species, is one such practice. Species are defined by their inability to interbreed or, more accurately, the failure of interspecies couplings to bring forth viable offspring that are capable of passing their own traits on to subsequent generations. This does not mean that interspecies crosses cannot result in progeny, only that the progeny is not reproductively competent. For example, a mule[4] – the result of crossing a female horse *Equus ferus caballus* with a male donkey *Equus africanus asinus* – has an odd number of chromosomes (63, a combination of 32 maternal chromosomes and 31 paternal chromosome) and is usually sterile as a result. Other than this reproductive dysfunctionality, mules are widely considered to be superior stock, as they tend to inherit the size and speed of their mothers and the endurance, sure-footedness, and disposition of their fathers. Charles Darwin himself was reportedly flummoxed by the "best of both worlds" attributes of the mule, writing, "That the offspring of the horse and the ass should possess more reason, memory, obstinacy, social affection, powers of muscular endurance, and length of life, than either of its parents, seems to indicate that art has here outdone nature" (Darwin, 1879: 34).

The mechanism by which mules exceed the fitness of both horses and donkeys, known as "hybrid vigor" or "heterosis", is not an exclusively equine phenomenon. Not all hybridizations result in improved offspring, but a number of them do. For example, various crosses between diadromous striped bass and freshwater white bass, white perch, and yellow bass – collectively known as hybrid striped basses – exhibit hybrid vigor, to a greater or lesser extent. The most popular of these hybrids, sunshine bass (female white bass × male striped bass) and palmetto bass (female striped bass × male white bass), are strongly heterotic and, as a result, commonly raised as food fish and for natural resource enhancement and sport fisheries (Harrell et al., 1990). Other notable hybrid fish include saugeye (walleye × sauger); cutbow (rainbow trout × cutthroat trout), splake (lake trout × brook trout), and various other interspecies crosses of *Oncorhynchus*, *Salvelinus*, and *Salmo* spp.; numerous crosses within the centrarchid genera *Lepomis, Pomoxis, and Micropterus*;

---

[4] "Mule" can also refer to a popular style of women's shoe. Backless, but typically with a closed toe, the mule is also a hybrid of sorts. However, the style of shoe does not share its etymology with offspring of male donkeys and female horses: the footwear takes its name from the French for slipper, whereas the term for the equine hybrid is thought to have Germanic origin.

and various ictalurid catfish hybrids, just among North American freshwater taxa (Wagner and Oplinger, 2013). In some cases, the offspring cannot reproduce, but in many cases they are at least marginally fertile, underscoring the fluidity of speciation and its oftentimes unfinished state.[5]

A great number of fish have been subjected to genetic improvement for aquaculture, including selective breeding, hybridization, and other manipulations to draw forth the desired traits (Hulata, 2011). Genetic engineering and direct transgenesis is just the most recent of these techniques to be investigated. Thanks to the high fecundity and external fertilization of most fish, reproductive manipulations that are unthinkable or near impossible in terrestrial vertebrates – including interspecies crosses, triploidy, and gynogenesis – are routine in aquaculture. So perhaps it is unsurprising that fish are at the forefront of GM food animals. In the late 1980s, Garth Fletcher and his colleagues at the Memorial University of Newfoundland were developing two lines of GM Atlantic salmon: one more tolerant of icy water temperatures because of an antifreeze protein gene construct taken from winter flounder and the other with exceptionally fast growth rates thanks to the insertion of a growth hormone-regulating gene construct based on genetic material from Chinook salmon and ocean pout (Fletcher and Hew, 2002). The AquAdvantage™ salmon hails from the latter line. Normally, Atlantic salmon grow fastest in the spring and summer months when water temperatures are warmest. Thanks to the Chinook salmon gene and the ocean pout promoters to encourage expression, the AquAdvantage fish produce growth hormone regardless of water temperature. As a result, AquAdvantage fish grow rapidly year-round and reportedly reach a marketable size in about half the time that it takes a normal Atlantic salmon.

---

[5] In *River Out of Eden* (1995), Richard Dawkins eloquently described the process of speciation as streams meandering outside of their banks: if separated long enough, divergent streams will cut banks too deep for their flood waters to overtop, but when the channels are only tentatively scoured, streams may diverge and reconverge easily. Such is the nature of speciation and the ability of organisms to commingle their genes.

The AquAdvantage salmon is but one of many transgenic fish in development for commercial aquaculture. AquaBounty Technologies has developed GM lines of tilapia, trout, and shrimp, and there are reportedly more than three dozen other GM fish in development throughout the world (Bailey, 2015). In addition to techniques focused on inducing over- or continuous expression of growth hormone, there is interest in using transgenic techniques to induce "double muscling" or the creation of a secondary layer of muscle tissue,[6] enhance tolerance of altered environmental conditions, or improve immunity and disease resistance. Transgenic poultry, swine, and cattle have also been developed for analogous purposes of improving growth, efficiency, immunocompetence, and so on (Forabosco et al., 2013).

Ostensibly, each of these genetic modifications and improved traits would be beneficial in commercial aquaculture and livestock production, but economic advantage is not the only consideration relevant to the debate over transgenic animals. An early review of GMOs in aquaculture outlined a number of general considerations and requirements for the production of transgenic fish in commercial aquaculture (Male et al., 1993). These include the need for the gene construct – the gene sequence itself and the associated promoters/enhancers – to be well-understood, stable, detectable in the recipient organism, safe, and commercially relevant. The review also highlighted technological requirements, such as the ability to induce gene expression in the appropriate tissue or organ and produce characteristics that are heritable. Other requirements were of a more regulatory or philosophical nature, including the need to understand the repercussions of GMO escapement and interbreeding with wild fish and to appreciate the ethical implications of transgenesis. It is the latter issue that is perhaps the most essential, but difficult

---

[6] Though conceptually bizarre to those unfamiliar with modern livestock production, double muscling is not foreign to animal science: Belgian blue, Piedmontese, and Parthenais cattle are all double-muscled breeds. Each possess a naturally occurring genetic mutation that suppresses production of myostatin, resulting in less restricted muscle growth and the hypermuscular build of these breeds.

to resolve. One could argue that the cumulative effect of selective breeding or hybridization in transforming the molecular and physical identity of a species is more profound than, for instance, the effect of adding a single Chinook salmon growth hormone-regulating gene into the genome of Atlantic salmon. The effect of this gene insertion is potent, multiplying the growth rate of these fish several times over, but is the act of splicing a single Chinook salmon gene into the Atlantic salmon genome of 37,206 genes (Lien et al., 2016) more invasive or ethically problematic than fertilizing the eggs of one salmonid with milt from another? Is a transgenic fish any more morally problematic than a mule or a GM soybean?

In late 2015, the U.S. Food and Drug Administration approved a new animal drug application for AquAdvantage salmon.[7] The agency concluded that, under the conditions of the approval, the fish posed no threat to consumers or the environment and could be marketed in the United States. Included among these conditions is the prohibition against rearing of AquAdvantage salmon in net pens to avoid the possibility of escapement. Health Canada issued a similar approval in May 2016 (Health Canada, 2016a). Despite both approvals relying on stringent environmental protection requirements, at this time, there is no requirement to label AquAdvantage salmon as a GMO when sold to consumers in the United States or Canada (Health Canada, 2016b; U.S. Food and Drug Administration, 2015e). This news unnerved a good number of Americans, Canadians, and others concerned about GM fish finding their way into their shopping basket. Of course, AquAdvantage salmon is far from the only GM food that may or may not have been in their food supply: there are 134 crop lines approved in Canada for direct use as foods or food additives; in the United States, there are 182. GM apples, melons, papayas, potatoes, corn, soybeans,

---

[7] In the United States, GM crops are subject to Department of Agriculture oversight, but GM animals are regulated by the Food and Drug Administration. The transferred genetic material (Chinook salmon gene and ocean pout promoter, in the case of AquAdvantage salmon) is considered an animal drug, and therefore falls within the agency's purview. See Chapter 7 for additional information about the aquatic animal drug approval process in the United States.

beets ... and now Atlantic salmon. Whether GM fish become commonplace in these countries or elsewhere in the world remains to be seen. GM fish deserve and will no doubt receive further scrutiny, but ideally no more or less than other GM foods.

Chapter 9

Food handling, safety, and quality

Guests, like fish, begin to smell after three days. (Benjamin Franklin, Poor Richard's Almanack)

Seafood is one of the most perishable foods, and improperly handled, prepared, or stored fish and shellfish can prove to be an ill-advised, even deadly meal. Perhaps it is for this reason that we have such a visceral, negative reaction to spoiled seafood – the pungent smell of fish that has gone off has undoubtedly kept many of our ancestors from poisoning themselves. There are few things as aversive as smelly seafood – just one whiff and we know something is wrong.[1] Fish and shellfish allergies are quite common[2] and the rare, but severe consequences of mercury-contaminated seafood are described in Chapter 6, but there are a number of other ways in

---

[1] This may explain why we say that something "smells fishy" when we suspect trickery or other malfeasance is afoot.

[2] It is estimated that roughly 2% of Americans are allergic to fish, shellfish, or both. Some allergies present as mild skin irritation whereas other more severe reactions may require epinephrine administration and/or hospitalization after ingestion or other contact with seafood. Shellfish allergies are approximately twice as common as fish allergies, but a variety of common seafoods have been associated with allergic reactions, including salmon, tuna, catfish, cod, shrimp, crab, and lobster (Sicherer et al., 2004). Along with peanuts, bovine milk, and eggs, seafood is one of the most common food allergy triggers observed worldwide (Prescott et al., 2013), and is particularly prevalent in regions where seafood is a dietary staple (Ryder et al., 2014).

which seafood can make us sick (Figure 9.1). Eating seafood can also result in physical injury via bone or shell fragments or other debris not removed during processing, though this aspect of food safety is hardly unique to seafood. Some of these risk factors are more commonly associated with wild-caught fish (ciguatera poisoning), whereas others are just as likely to affect farmed seafood (histamine poisoning).

In addition to being one of the most perishable foods, fish and shellfish have become one of the most widely traded foods: the value of international trade of seafood in 1976 was US$8 billion; today, the value exceeds $102 billion (Ryder et al., 2014). As our seafood supply navigates the globe,[3] it encounters different approaches and standards for food handling, and every link in the supply chain is a possible contributor to declining quality. "Quality" can be a subjective term, but when used to describe seafood it has certain, specific meaning: in this context, quality speaks to the composition and food safety attributes of seafood, and implies that it will meet consumers' expectations that food be wholesome, nutritious, safe, hygienically produced, within its expected shelf-life, and have product-specific sensory characteristics (Ryder et al., 2014). In this chapter, we examine the various elements of seafood quality, gauge the relative risks in the fishing and aquaculture sectors, and ask whether any of this registers in the collective consciousness of consumers.

Although the phrase "A fish rots from the head"[4] would suggest otherwise, the process of decomposition typically begins in the gut. Multitudes of bacteria, no longer kept in check by the fish's

---

[3] The United States imports more seafood than nearly every other country on Earth – roughly 90% of the seafood Americans eat comes from somewhere else. The average straight-line distance seafood travels to reach them is 8812 km; depending on where the seafood is imported from, the distance can approach nearly twice that figure (McClenachan et al., 2014). On average, this is equivalent to seafood in Los Angeles traveling to New York City and back before being eaten.

[4] This folksy idiom is known in various forms in Turkey, Russia, and elsewhere throughout the world, but appears to be – like so much else – Greek in origin. Meaning that misbehavior or failures in a household, business, or other group can be traced to the actions (or inactions) of leadership, the phrase is more illustrative than biologically accurate (Anon., 1988).

Toxins	Ciguatera poisoning	Toxins produced by a dinoflagellate accumulate in warmwater marine fish, tissues, most commonly in larger reef fish such as moray eels, barracuda, red snapper, and amberjack
	Scombroid or histamine poisoning	Histamine accumulates in the tissues of improperly handled or stored fish, most commonly mackerel and tuna for which the disease is named, as well as amberjack, mahi-mahi, and swordfish
	Pufferfish or tetrodotoxin poisoning	Caused by ingestion of tissues containing or contaminated by improper handling with tetrodotoxin, most commonly pufferfish but also including certain octopi; amphibians, echinoderms, and horseshoe crabs
	Shellfish poisoning (SP) syndromes	Toxins produced various algae (saxitoxin = paralytic SP, domoic acid = amnesic SP, okadaic acid = diarrhetic SP, and brevetoxins = neurotoxic SP) are concentrated by the filter-feeding activity of shellfish in waters contaminated by harmful algal blooms, typically clams, oysters, mussels, and other bivalves

Figure 9.1: The most common causes of seafood-borne illness in humans (Gresham and Taylor, 2015; Huss, 1994; Iwamoto et al., 2010) include toxins produced by the aquatic animals themselves or microorganisms in the environments they inhabit as well as a range of infectious agents. In most cases, disease is associated with unsanitary or inhospitable environmental conditions (for example, waters contaminated by sewage discharge or harmful algal blooms) or improper handling and storage that allow otherwise wholesome foods to become contaminated or go bad. Although finfish were to blame for the majority of seafood-borne illness in the United States in the 1970s and 1980s, more outbreaks in the 1990s and early to mid-2000s were caused by consumption of mollusks (>55%), particularly raw oysters and other bivalves, than crustaceans (<20%) or finfish (<25%) (Iwamoto et al., 2010). However, finfish-related illness may be on the rise: in 2013, fish were responsible for 24% of the foodborne-related disease outbreaks in the United States – more than any other food type and more than double the number of outbreaks associated with mollusks (Centers for Disease Control and Prevention, 2015b). Every year, approximately 260,000 Americans suffer food poisoning after eating finfish, and of the outbreaks associated with finfish from 1998–2015, the vast majority of cases were caused by scombrotoxin (55%) and ciguatoxin (36%), mostly associated with consumption of tunas, mahi-mahi or common dolphinfish, and groupers. Although this study did not address the origin of the implicated fishes, none of these species are routinely cultivated, suggesting that wild fish are responsible for the majority of reported cases (Barrett et al., 2017).

Pathogens	Bacteria	*Vibrio* spp., disease most commonly associated with consumption of undercooked seafood, particularly raw oysters or other bivalves that accumulate the bacteria by filter feeding
		Salmonella spp., disease most commonly associated with consumption of undercooked seafood harvested or cultured in sewage-contaminated waters or contaminated by improper handling or preparation
		Shigella spp., disease most commonly associated with consumption of undercooked seafood harvested or cultured in sewage-contaminated waters or contaminated by improper handling or preparation
		Clostridium botulinum, spores found naturally in various aquatic animal tissues, but disease is most commonly associated with improperly fermented foods and traditional Alaskan Native foods prepared from salmon heads and eggs and marine mammals
		Other bacteria, including *Staphylococcus, Clostridium, Listeria*, and *Bacillus* spp., typically associated with improper handling and storage of seafood
	Viruses	Norovirus, disease most commonly associated with consumption of undercooked seafood, particularly filter-feeding bivalves, harvested or cultured in sewage-contaminated waters or contaminated by improper handling or preparation
		Hepatitis A virus, disease most commonly associated with consumption of undercooked seafood, particularly filter-feeding bivalves, harvested or cultured in sewage-contaminated waters or contaminated by improper handling or preparation
	Parasites	Helminths (e.g., *Diphyllobothrium* spp. and Anisakdae) are common parasites of freshwater and marine finfish, disease most commonly associated with consumption of never frozen, undercooked seafood, such as sushi or raw shellfish
		Seafood-associated protozoa causing disease in humans are largely water-borne (e.g., *Giardia* spp.), meaning that disease is rarely caused by the seafood itself, but by the water from which the seafood was harvested

Figure 9.1: (Continued)

bodily functions, proliferate and begin to break down the surrounding tissues in search of water and other resources. As these bacteria work from the inside out they are joined by others on the surface working in the opposite direction. Some may be associated with foodborne illnesses; other microbes are relatively benign, but rapidly alter the texture, aroma, and taste of seafood eventually rendering it inedible. Would-be decomposers are influenced by their environment, and the timeline of decomposition depends on temperature and the availability of water, oxygen, acidity, and other variables. For example, *Listeria monocytogenes* (the bacterium responsible for listeriosis), does not grow at temperatures below −2°C or above 45°C, in very acidic (<4.2 pH) or basic (>9.5) conditions, when moisture availability is less than about 1%, or in the presence of nitrite or carbon dioxide at sufficiently high concentrations. Spoilage is also affected by a number of other processes that can speed rancidity, such as "self-digestion" by the body's own enzymes (autolysis), or chemical degradation due to the presence of oxygen (oxidation) or water (hydrolysis). Like microbial decomposition, these processes are influenced by ambient temperature and other environmental conditions. Proper food handling and storage strategies target these variables in the attempt to slow or arrest microbiological activity and the physical processes of spoilage (Ryder et al., 2014).

What an animal experiences in the moments before it is dispatched can dramatically influence the quality of the resulting meat,[5] but the biological and chemical progression to spoilage does not begin until after death. Delaying slaughter as long as possible is an obvious means of forestalling spoilage, and indeed this may have been one

---

[5] Animals experiencing physical or psychological stress prior to slaughter typically yield less due to losses in processing efficiency, and product quality is often reduced by a variety of physiological factors affecting meat texture, shelf-life, and so on. Temple Grandin, regarded primarily as an advocate for the humane treatment of livestock, often argued for changes to rearing and slaughter methods from the position of improving product quality, not animal welfare for its own sake. Improving conditions and reducing preslaughter stress can also improve fillet quality in fish (Trushenski et al., 2017).

of the early motivations for holding and cultivating fish near the home (see Chapter 1). In the modern era, aquaculture still offers the opportunity to harvest fish more or less on-demand and live-holds allow wild catches to be kept alive until the fishing fleet returns to port. The data are mixed based on species and harvest and handling circumstances, but rested harvest (fish are sedated prior to slaughter) and rapid slaughter methods (electrocution, percussion, and the like) tend to give superior results. Post-mortem, preserving quality is about reducing temperature as quickly as possible, hygienic handling and processing, and altering the seafood "environment" so that it is less favorable to microbial growth and biochemical spoilage. Washing removes debris and reduces the abundance of microbes on the skin or shell. Rapidly reducing temperatures with chilled water, ice, or mechanical refrigeration slows microbial growth and autolysis. Bleeding and gutting reduces the amount of water, oxygen, and microbial load present in the body, and facilitates rapid cooling of large-bodied individuals. Washing after evisceration is important to clean the body cavity and surfaces prior to filleting or other processing (Borderias and Sanchez-Alonso, 2011). After processing, seafood can be refrigerated, frozen, canned, or vacuum-packed; dried or smoked; salted, marinated, or fermented; or treated with artificial preservatives. In many cases, these techniques are used in combination to extend shelf-life from a few days to a few years or longer (Ryder et al., 2014).

Although there is considerable overlap in the handling, processing, and preservation of both farmed and wild fish, greater predictability and control over product quality and closer proximity to processing facilities tends to favor farmed seafood in terms of product safety and quality (Borderias and Sanchez-Alonso, 2011; Rubino, 2008). As noted above, minimizing stress and the amount of time between capture and death generally improves seafood quality, as does purging fish prior to slaughter and processing. Although such strategies are readily implemented in an aquaculture setting, there is often little that can be done to address these factors in industrialized fishing operations (Taklemariam et al., 2015). The (in)ability to control where and how fish are reared, caught and killed, and so forth can translate into differences in the quality of farmed versus wild seafood products. For example, farmed Nile tilapia fillets had significantly lower pH than wild-caught fillets after slaughter, but higher concentrations of total volatile nitrogen, indicating more advanced spoilage in the wild-caught fish (Baz et al., 2014). In a recent review of various aspects of aquatic food security in the United Kingdom, it was noted that wild and farmed seafood present essentially the same biological (infectious agents) and chemical (heavy metals and other contaminants) health risks to consumers, but that there are a number of opportunities to mitigate or eliminate these threats for farmed, but not wild seafood supplies (Jennings et al., 2016).

And what of the culinary quality of farmed versus wild seafood? Beyond the nutritional quality and safety issues addressed elsewhere in this title (see Chapters 5, 6, and 7), there is the matter of taste, aroma, texture, and the other sensory characteristics of seafood. Some people eat seafood simply to satisfy their bodies' need for protein and other nutrients. Others intentionally pile their plates with fish and shellfish with the benefits of long-chain omega-3 fatty acids in mind. Of course, there are a good number of us who are not so dietetically diligent or concerned with the nutrients provided by seafood or anything else we might eat. We eat seafood to savor the clean, briny sweetness of an oyster … the exquisite melting of toro on the tongue … the crunchy, meaty goodness of fish and chips. For those of us who eat seafood for the sheer pleasure of it, how do farmed fish stack up against their wild counterparts? As it turns out, much of what separates the culinary experience of eating farmed versus wild seafood may be in our heads, not our mouths. When farmed and wild fish are compared in the cold, objective light of the food science laboratory, the shadows of what consumers think they know about the appearance and taste of farmed versus wild seafood disappear. What remains is often the banality of near-equivalence. Farmed fish tend to contain more fat – and therefore more beneficial fatty acids and fat-soluble vitamins – than wild fish, but are often quite comparable otherwise. Sometimes food scientists are able to demonstrate differences in the color, texture, taste, and smell of farmed versus wild-caught seafood, but differences that are readily apparent in the laboratory are not always obvious in the kitchen or at the dinner table. The average consumer is often unable to differentiate between farmed and wild fish. For example, a trained taste test panel was able to describe differences in the appearance and taste of farmed and wild Atlantic cod products,[6] but these differences did not appear to be strongly related to consumer preferences, as there were groups that preferred the wild cod and others that preferred the farmed fish (Sveinsdottir et al., 2009). Similarly, consumers found the taste, aroma, and texture of farmed and wild-caught tilapias to be largely equivalent (Joram and Kapute, 2016).

Although most consumers are not especially knowledgeable about the nature of the seafood they consume, most tend to perceive aquaculture products as cheaper and more available, but less wholesome, nutritious, and of a lower quality compared to wild-caught fish and shellfish (Claret et al., 2014; Pieniak et al., 2013). Others may avoid farmed seafood because they are unsure about its country of origin or sustainability (Conte et al., 2014). As discussed elsewhere in this book, the objective reality is often quite different, and there is evidence that consumers' preferences are informed largely by the *idea* of wild versus farmed fish. When asked to describe the basis for their preferences for wild fish over farmed fish, Belgian consumers responded that "Wild fish grew up in a natural way; they were never forced," "Wild fish are happier; they can swim wherever they want, move freely, which makes the fish stronger," and "Farmed fish are less healthy, because of growing up in an artificial environment, and because of being given growth promoters and antibiotics" (Verbeke et al., 2007: 129). Most consumers know next to nothing about where seafood comes from and the realities of fishing and fish farming. They fill that void with emotions and misjudge aquaculture products they would otherwise find acceptable, even preferable. When consumers were asked to compare farmed and wild black spot sea bream, gilthead sea bream, sea bass, and turbot in side-by-side tests, most preferred the farmed fish … that is, unless they were told which fillets were which (Claret et al., 2016). Farmed yellow perch beat a variety of wild-caught white fish in blind taste tests (Delwiche

---

[6] The farmed cod products tended to have a lighter and more consistent coloration and sweeter, meatier flavor, but a more rubbery, less tender texture than most of the wild cod products (Sveinsdottir et al., 2009).

et al., 2006), but one wonders if the results would have been different had the participants known exactly what they were eating.

The perceived inferiority of farm-raised seafood likely has more to do with uninformed consumer attitudes than the products themselves. A survey of consumers from Belgium, Denmark, the Netherlands, Poland, and Spain revealed that the public is not especially knowledgeable about the seafood they consume: although the majority of respondents correctly identified salmon as a fatty fish (55–83%), far fewer knew that one-half of the fish available to them at market were farm-raised (31–56%) (Pieniak et al., 2006). Consumers in Germany and the United Kingdom were also found to be ill-informed, confused, and skeptical of aquaculture and organic salmon farming in particular (Aarste et al., 2004); Greek consumers were similarly uninformed, but reluctant to purchase farmed fish (Arvanitoyannis et al., 2004). Although consumers appear to prefer wild-caught seafood in much of the world, these findings are far from universal. More recent evaluations of the German and Spanish seafood markets reaffirmed the overall preference for wild-caught over farmed fish, but also documented the emergence of a subset of consumers that prefer aquaculture products (Bronnmann and Asche, 2017; Claret et al., 2012). Farmed and wild-caught salmon appear equally competitive in Japanese markets (Asche et al., 2003) and, in some circumstances, consumers in the Northeastern and Mid-Atlantic United States expressed a preference for farmed oysters (Kecinski et al., 2017) and farmed salmon based on assumptions that the farmed product is safer and of a higher quality than wild salmon (Holland and Wessells, 1998). In the latter case, the preference for farmed salmon only emerged when the products were priced similarly, and consumers were assured that all of the salmon they were presented with had passed a safety inspection. More so than questions of farmed versus wild, lingering concerns over perceived safety and the all-important influence of price appear to weigh most heavily in consumers' minds.

Norwegian consumers were not troubled by media coverage of toxic residues in farmed versus wild fish and were generally trusting of their government's control officials' ability to protect them from foodborne illness, but some expressed a preference for wild salmon based on their taste, lower fat content, and the less tangible notion of "wildness" being superior to domestication. When German consumers were similarly interviewed, they expressed a much more skeptical view of food safety control measures and of aquaculture, citing concerns over contaminants in farmed fish and the ethics and environmental consequences of intensive aquaculture and the perceived purity and wholesomeness of wild fish. That said, seafood producers and brokers noted that many consumers in both countries have grown accustomed to the consistency of farmed salmon and reject the more variable wild product when it does not meet their expectations of quality and color, suggesting that the stated preferences of consumers may not be fully reflected in their purchases (Dulsrud et al., 2006). Stated preferences for wild fish have also been reported for consumers elsewhere in Europe, Canada, Ghana, Kenya, Egypt, Tanzania, and the United States (Bacher, 2015; Darko et al., 2016; Davidson et al., 2012), which may or may not agree with their purchasing habits: in Portugal and British Columbia, Canada, the strong stated preference for wild-caught seafood was in direct opposition with widespread consumption of aquaculture products (Cardoso et al., 2013; Murray et al., 2017). Unless consumers have experienced food safety threats first-hand or purchased products that turned out to be spoiled or otherwise inferior, seafood selection and purchases tend to be influenced by little other than price (Bacher, 2015).

Traceability – the ability to track food products back to their point of origin – is an essential element of food safety in that it allows product recalls to be prosecuted quickly before most of the identified product can reach consumers. Traceability is a known difficulty for most seafoods. The first challenge may be in verifying that the fish is the species it is purported to be, a surprisingly difficult task in some cases. The ability to track products is further complicated by the many countries of origin, supply routes that suggest webs rather than chains, and the fact that the same product may hail from a farm or wild catch (Moretti et al., 2003). A simulated recall in Norway revealed that 40% of seafood products could not be traced, even after dozens of phone calls, emails, and other attempts to track the products back to their sources. This study did not suggest any major differences in the traceability of farmed versus wild-caught seafood (Karlsen and Senneset, 2006). Although solutions to these problems are far from simple to resolve, at least for aquaculture, the growing popularity of third-party certification programs that require traceability provisions (Food and Agriculture Organization, 2011) will help to improve the efficiency of tracking products at any moment in time and tracing them to their point of origin.

When we eat, we seek to feed both body and soul. We hunger for the nutrients that comprise our physical selves, but we also crave foods that express who we are in a more philosophical sense. The foods that define our diets – what we eat and what we deem taboo – reflect religion and culture. We commemorate events and remember the places and people we love by preparing traditional meals. We cook "comfort foods" to soothe our minds as much as our bellies. Strangers become friends across the dinner table, and we affirm long-standing relationships by sharing meals. It has been said that we taste with the eyes before the tongue,[7] but perhaps all meals begin earlier in our hearts and minds. Most consumers persist in the belief that aquaculture products cannot feed their bodies or souls as wild fish and shellfish can, that farmed seafood does not reflect their tastes. Farmed seafood is as safe, wholesome, nutritious, and pleasing to the palate as wild-caught seafood, but we must accept the invitation to dine – in terms of philosophy, ethics, and identity – before we can acquire the taste for aquaculture and enjoy the benefits and pleasures of eating farmed seafood.

---

[7] A good number of cooks have undoubtedly uttered similar phrases, but "The first taste is with the eyes" is typically attributed to Apicius, a Roman gourmand alive during the 1st century CE.

PART III

Environmental issues

Chapter 10

Production systems and water usage

> Water's role in underpinning all aspects of sustainable development has become widely recognized. It is now universally accepted that water is an essential primary natural resource upon which nearly all social and economic activities and ecosystem functions depend. (UNESCO, 2015: 9)

Few things are better than a tall glass of cold water on a hot day; few things are worse than an unslaked thirst. Thirst can drive a person mad, and conflicts over water can escalate to open war. Disconcertingly, fresh water is becoming an increasingly scarce commodity. In fact, thinking of water as a "commodity" at all is troubling. More than 70% of Earth's surface is covered with water, but more than 97% of the blue planet's water is saline – undrinkable and largely unusable for food production.[1] Less than 3% of Earth's water is freshwater, and more than two-thirds of this is inaccessible, locked away in the perpetual snow and ice of the Arctic and Antarctic circles. That leaves less than 1% as liquid freshwater in surface waters and underground water supplies to satisfy the thirst of all life on Earth (U.S. Geological Survey, 2016a) (Figure 10.1). Access to a safe, adequate water supply is essential to life, biologically and in terms of quality of life: clean water is a prerequisite to food security, proper sanitation, healthcare, energy and economic development, and so on (UNESCO, 2015). A staggering 780 million people do

---

[1] Save for aquaculture, of course.

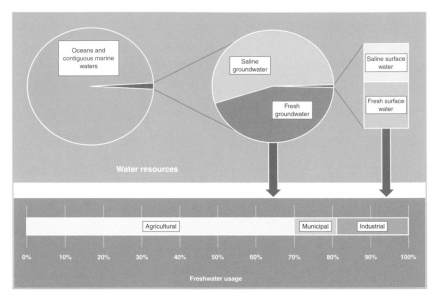

Figure 10.1: The oceans represent the vast majority of Earth's liquid water resources. Most of the remaining fresh- and saltwater is found underground, with the world's lake, rivers, and other surface waters only representing the slimmest fraction of global water sources. Less than 1% of all the water on Earth is liquid freshwater, and humans tap both groundwater and surface waters to fulfill demands for drinking water, sanitation, crop irrigation and livestock watering, and various industrial uses of freshwater.

Sources: U.S. Geological Survey (2016b) and UNESCO (2015).

not have access to clean drinking water, and this number will likely increase as mankind continues to multiply. Roughly 85% of the human population lives in the driest half of the planet (United Nations, n.d.), and population growth is particularly rapid in some of the most drought-prone regions in the world, putting an even greater strain on limited supplies. With population growth comes urbanization, greater food and fuel consumption, intensive agriculture and industrialization, and greater energy demand, all of which dramatically increase societal water needs. Because of growing affluence and the attendant increases in per capita caloric intake and meat consumption, use of motor vehicles, and other forms of energy use, demand for freshwater has increased at twice the rate of human population growth for decades. Climate change further complicates water supply and demand forecasts, but increasing temperatures, altered rainfall patterns, and more erratic, extreme weather offer little reason to be optimistic about the future of already scarce freshwater resources (UNESCO, 2015).

Agriculture is the single largest human use of freshwater. The United Nations pegs crop irrigation and livestock production at 70% of total freshwater use (Figure 10.1), but this figure may be an underestimate as it considers consumption of surface and groundwater only. Accounting for water use is complex: comprehensive assessments account for rainwater falling directly on cropland, groundwater extraction and surface water diversion, as well as water that is not used, *per se*, but is rendered unfit for other uses as a result of pollution. Accounting for all of these variables, agriculture may be an even bigger consumer of freshwater resources, responsible for more than 90% of humanity's water "footprint". Of the 1025 cubic gigameters $(Gm^3)^2$ of fresh ground- and surface water used annually throughout the world, municipal (42 Gm^3), industrial (38 Gm^3), and livestock (46 Gm^3) supplies are a veritable drop in the bucket compared to the remaining 899 Gm^3 used in crop irrigation (Hoekstra and Mekonnen, 2012). Regardless of the methods of accounting, most of the world's freshwater is used to produce food and none of the estimates leave much room for expansion of the agricultural sector.

Countries with large populations naturally have large water footprints, but not all countries use water equally per capita. China, India, and the United States are the first, second, and third-most populous countries on Earth and their water footprints are identically ranked, but whereas China and India are each home to roughly 18% of humanity, the United States can only claim just over 4% of the human population. Consequently, the United States uses more water per capita than most other countries on Earth (Hoekstra and Chapagain, 2007; Hoekstra and Mekonnen, 2012). More than 40% of the nation's freshwater use goes to food production, but it is mostly used to irrigate crops, not water livestock or raise fish

---

[2] One cubic gigameter is equivalent to 1,000,000,000,000,000,000,000,000,00 0,000 liters or 264,200,000,000,000,000,000,000,000,000 gallons.

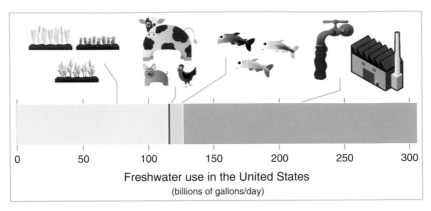

Figure 10.2: Every day, more than 300 billion gallons of freshwater are dedicated to agricultural, industrial, and municipal use in the U.S. Approximately 75% of the freshwater supply comes from diversion of surface waters and the rest is extracted from groundwater sources (U.S. Geological Survey, 2016b).

Source: Crops, livestock, and fish / User: Viscious-Speed / OpenClipArt / Public Domain / Creative Commons Zero 1.0 License, Faucet / User: j4p4n / OpenClipArt / Public Domain / Creative Commons Zero 1.0 License, Factory / User: rg1024 / OpenClipArt / Public Domain / Creative Commons Zero 1.0 License.

(Figure 10.2). In the United States, more surface and ground water is used for aquaculture than terrestrial livestock production: roughly 2 billion gallons of water are used daily to raise cattle, pigs, chickens, and the rest, whereas about 9.4 billion gallons are used to raise fish and shellfish.[3] These volumes may seem staggering, but livestock accounts for a little more than 0.6% of U.S. water withdrawals and aquaculture amounts to just over 3%; irrigation accounts for over 37% of freshwater use (U.S. Geological Survey, 2016b).

Intuitively, one would expect animals *living* in water to use more of it than land animals, but much of the water that is "used" to raise fish in the United States is directly returned to the environment. For example, at Idaho's trout farms – the top-ranked user of the nation's water in the aquaculture sector – water comes in and, after being cleaned up, goes right back out. To fully understand use

---

[3] For those more metrically inclined, daily freshwater use in United States aquaculture is roughly 35.7 billion liters and livestock watering amounts to 7.6 billion liters.

of water in aquaculture in the United States or anywhere else, one must understand the various types of rearing systems that are used to raise fish and shellfish. All aquaculture systems use water, but in different ways and quantities depending on whether the system is recreating the ecosystem services that keep aquatic organisms alive in nature (as in recirculation aquaculture systems), harnessing these functions (flow-through tanks, net pens, rafts, or cages), or doing a bit of both (ponds). All living organisms require water, but aquatic life – cultivated or otherwise – depends on water so completely there is no analogous medium for terrestrial life. Beyond the essential H_2O, water maintains ambient temperatures and sets the pace for all biochemical processes, provides oxygen and key electrolytes, serves as a sink for carbon dioxide, ammonia, and other metabolic byproducts, and so on – water is both part and parcel of the life support for feral and cultivated aquatic organisms.

Life is infinitely complex, but largely boils down to an interplay between oxygen, carbon, and nitrogen. Primary producers, like plants and algae, require carbon dioxide and nitrogen-bearing nitrate, nitrite, or ammonia as raw materials, and the environment must accommodate the excess oxygen they produce as a byproduct of photosynthesis. Conversely, fish, shellfish, and other aquatic animals must be provided with oxygen and a means of dissipating the ammonia and carbon dioxide they release as wastes. One way or another, aquaculture systems must address these elemental needs.

Floating rafts used to cultivate seaweeds or sedentary bivalves and net pens or cages used to culture finfish are wholly reliant on the environment to maintain water quality. They do not harness ecosystem services so much as they submit and are beholden to them. Such open systems do not consume water in a literal sense, but they do influence its composition by way of their biological functions. Carbon dioxide produced by finfish and shellfish cultured this way is largely consumed by phytoplankton and other local photosynthesizers that provide an ample supply of oxygen in return. Omnipresent nitrifying bacteria transform discharged ammonia into nitrite and nitrate that are less toxic and more readily absorbed by aquatic plants and algae. Outputs from finfish systems include these dissolved materials and solid wastes, such uneaten feed and fecal matter, that settle immediately to the benthos or are flushed by currents and deposited elsewhere. The cumulative impact of discharges from fish cages or net pens depends on the volume of solid and dissolved waste as well as the composition and carrying capacity of the receiving system. In the case of filter-feeding organisms, open systems are net consumers of solids and water clarity typically *increases* as cultivated bivalves or fish hungrily harvest particulate foods from the water. Discharge of waste from open cultures of seaweed to the environment is similarly "net negative", and these operations release little other than the oxygen they generate. Regardless of the type of organism being cultivated, open aquaculture systems are directly and completely connected to the ecosystems of which they are a part, and each would fail if not for the currents, tidal flux, nitrogen cycling, and photosynthesizers that remove wastes and provide oxygen and other life-giving nutrients. In the United States, large net pen operations are designated as Concentrated Aquatic Animal Production (CAAP) facilities and are subject to regulation by the U.S. Environmental Protection Agency (USEPA) under the National Pollutant Discharge Elimination System (U.S. Environmental Protection Agency, 2017a). To operate, proprietors must secure a permit and monitor the surrounding water and sediments to ensure discharges from the open system do not contain unacceptable levels of nitrogen, phosphorus, solids, or other regulated materials. If conditions fall outside of those spelled out on the permit, the farm can be fined and the permit might be revoked if the operators do not alter production or implement some other means of addressing the amount of pollution generated by the farm.[4]

---

[4] NPDES permit limits are set to protect the quality of public waters. Facilities meeting the production thresholds to be considered CAAP facilities are subject to USEPA oversight, but even smaller operations may require NPDES permits. To stay compliant, facilities either limit production densities or direct effluents through on-site settling or treatment systems. Of the 96 permitted aquaculture facilities in the USEPA Enforcement and Compliance History Online Database, only two are currently listed as having significantly violated the conditions of their permit (U.S. Environmental Protection Agency, 2017b).

Flow-through systems are not situated in natural ecosystems like open systems are, but they are still tethered to the natural world by the water that moves through them, providing needed oxygen and sweeping away metabolic detritus. Flow-through systems do not use water consumptively, but water quality is altered after passing through one or more batteries of fish in tanks or raceways prior to release. Flow-through systems are designed to rely on influent and effluent waters to satisfy the environmental needs of cultured fish, while giving culturists the opportunity to adjust or amend conditions to improve performance. For example, an incoming water source may contain enough oxygen to pass through two raceways of fish in series without becoming dangerously depleted, but the culturist may wish to aerate the water before it is passed to a third raceway to ensure fish at the end of the line are not stressed by low dissolved oxygen levels. Alternatively, the culturist might choose to stock the third raceway with a species or life stage of fish that is more tolerant of lower oxygen. In terms of effluent management, flow-through systems also allow for water to be treated to remove or reduce levels of dissolved or solid wastes prior to discharge. Large flow-through systems in the United States are subject to regulation as CAAP facilities, and must monitor their effluents to ensure compliance with their discharge permits or face fines or other regulatory action by the USEPA.

Ponds are deceptively simple: an earthen depression is filled with water, and fish, shrimp, or other aquatic organisms are added. If there is to be a bountiful harvest, culturists must establish a delicate balance between inputs and outputs to maintain conditions that support a thriving aquatic community. Ponds are mesocosms in which water and nitrogen cycles, photosynthesis, respiration, and other natural processes must be carefully managed to ensure harmony and productivity. Generally speaking, water use in pond aquaculture is consumptive: other than evaporative losses and whatever volume might seep into the surrounding ground, water goes in and stays in. Over time, the cultured fish or shrimp will deplete oxygen levels and foul the water with their wastes. Unlike open- or flow-through systems that can rely on the greater capacity and environmental services of the ecosystem at large, ponds are relatively contained and must be self-sufficient in attending to the environmental needs of cultured aquatic organisms. As described in Chapter 3, ponds can be managed extensively – with few inputs – or intensively. Extensive aquaculture ponds rely on natural blooms of phytoplankton and rooted vegetation to balance the system's respiration debt and benthic sedimentation to integrate solid wastes. As production intensifies and the natural carrying capacity of the pond is approached, aeration, water exchanges, or other interventions are needed to maintain water quality conditions. Water is rarely discharged from aquaculture ponds, other than when ponds are drained to facilitate harvest[5] or to rework the pond bottoms and manage the nutrient-rich sediments.

Recirculation aquaculture systems (RAS) are the most technologically advanced, complex rearing systems. RAS are also sometimes referred to as water reuse systems because they are designed to do just that: water that has been fouled by the cultured organisms is treated to remove or neutralize excretions and other wastes, and so purified is reused. Ironically, these complex aquatic life support systems are probably the most familiar to hobbyists, for a home aquarium is nothing but a RAS in miniature. Whereas home aquaria might use continuously bubbled air to provide oxygen and circulate water through an under-gravel solids-cum-biological filter to remove uneaten feed, feces, and dissolved wastes, commercial RAS use different but analogous technologies and larger-scaled components to maintain water quality. In both cases, little-to-no water is exchanged during normal operation; freshwater is added only to compensate for evaporation or whatever water is lost during maintenance of system components such as when filters are cleared of debris. RAS have some obvious advantages over other culture systems, namely their very small water footprint and physical separation from the environment. Clean water, oxygen, food, anything needed by the cultured organisms is provided by the system's machinations or the

---

[5] Finfish are regularly harvested from ponds by seining – dragging the entire pond with a large net – but drawing water levels down makes this process easier. Some species, like shrimp, are more difficult to capture with a seine, so these ponds are often drained to ensure a complete harvest.

culturist. Because RAS do not rely on natural ecosystem services and are not functionally connected to the environment, they can be located anywhere[6]. This makes maintaining biosecurity a snap, and means that RAS have limited direct effects on the environment. Of course, there are reasons that RAS have not yet come to dominate aquaculture worldwide, principally the cost of constructing RAS and their exorbitant energy demands.[7] Like ponds, RAS concentrate wastes, and when discharges are periodically released they can be more problematic to handle than the dilute, continuous effluents of open or flow-through aquaculture systems.

It is incredibly difficult to compare water use across different forms of aquaculture, much less different forms of agriculture. Open- and flow-through aquaculture systems do not use water consumptively in the same way that ponds and RAS do, but "borrow" much greater volumes to operate. Less water is used to rear animals in RAS, but these low, direct-use figures mask the amount of water used in sanitation systems to mitigate the concentrated waste they generate. Natural foods providing nutrition to fish and shrimp in extensively managed ponds do not increase the water footprint of these operations, but feed inputs – and the agricultural products they contain – represent a significant, if indirect use of water in other forms of aquaculture. Analytical frameworks (life cycle analysis, water footprint) and methodological details vary substantially from study to

---

[6] Hypothetically, RAS can be operated anywhere on Earth or beyond. With the help of the Japanese Space Exploration Agency, RAS "slipped the surly bonds of Earth" (Magee, 1942) and have been used to rear a variety of finfish species on-board the International Space Station (National Aeronautics and Space Administration, 2012). Although these fish are in space to help researchers understand bone density loss and other effects of low gravity on the human body, the U.S. National Aeronautics and Space Administration has provided grants to researchers interested in developing space-ready RAS and aquaponics systems to grow food during space exploration (Martin, 1998).

[7] Advances in system design and operation have increased the energy efficiency of RAS such that they are approaching equivalence with some types of flow-through systems and capture fisheries. Growth in this sector is nonetheless slow, but use of RAS appears to be on the rise in Europe (Martins et al., 2010).

study and can yield "wildly different results" (Doreau et al., 2012).[8] Recognizing that much of the accounting work done to understand global water use is confounded by "apples-to-oranges" comparisons, several consistent truths appear to have emerged from this field of study. First, water use per unit of food tends to be lower for crops than livestock (Doreau et al., 2012; Hoekstra and Chapagain, 2007). Given the superior conversion efficiencies of aquatic animals (see Chapter 14), it stands to reason that using agricultural crops to feed fish instead of terrestrial livestock would increase the amount of animal protein produced within any particular feed-related water footprint. Second, per unit freshwater use tends to be lower for aquatic livestock than terrestrial livestock (Figure 10.3). Taken together, these two broad conclusions would suggest that aquaculture puts less of a strain on freshwater resources than other forms of animal agriculture. As freshwater becomes an over-allocated and scarce resource, aquaculture will have an even greater role to play in sustainable development (Federoff et al., 2010; Molden et al., 2010; Subasignhe et al., 2009).

There are also a number of ways to coax more food from limited water resources in aquaculture. Intensification is one obvious approach, but with it comes greater energy consumption and waste generation that are generally counter to the sustainability-minded interest in reducing water consumption. With respect to pond culture, the solution may be in the mix of nutrients, surface area, and species (Bosma and Verdegem, 2011). Selecting fertilizer inputs to minimize nitrogen and maximize carbon inputs can stimulate growth of film- and mat-forming microbial assemblages. The so-called "bioflocs" they form can provide a secondary source of nutrition for farmed fish and shrimp, improve water quality, and decrease nutrient loss in effluents or sediments. Increasing the surface area available to these and other beneficial microbes by adding

---

[8] For instance, life cycle analysis – which does not include rainfall or estimates of water "use" associated with pollution – suggests that 3–540 liters of water may be used to raise 1 kilogram of beef. Water footprint analysis – which does take some of these other freshwater types into account – suggests that it may take 10,000–200,000 liters of water to produce the same kilogram of beef (Doreau et al., 2012).

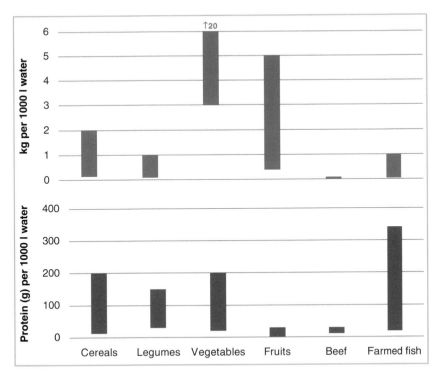

Figure 10.3: There are numerous ways to measure agricultural productivity per unit volume of freshwater use, including the crude weight of food (upper panel) and protein produced (lower panel). In either case, productivity varies among commodities, but is generally lower for terrestrial livestock than crops. General and protein productivity is substantially greater in aquaculture than in beef production and plant-based foods. Animal protein is generally more digestible and better matched with human dietary needs than plant protein, so aquaculture is well-positioned to provide a hungry world with a greater quantity and quality of protein (Molden et al., 2010).

submerged substrates (large woody debris, mesh-style plastic fencing material, and so forth) can further improve nutrient cycling in ponds and increase the abundance of natural foods (Bosma and Verdegem, 2011). Polyculture – cultivating multiple species within a single culture system – can increase production densities without substantially increasing the inputs needed to operate. Carp polyculture, with or without the addition of shrimp or other shellfish, is a tried and true method of increasing pond productivity and is practiced throughout much of Central and Eastern Europe and Asia.

European, Asian, and Indian major carps occupy different niches in their native ecosystems. For example, common carp are bottom-feeding detritivores, bighead carp and silver carp are filter-feeding planktivores, and grass carp feed primarily on aquatic vegetation. Fertilizing a pond will initially stimulate phytoplankton and zoo-plankton blooms, increasing the food base for bighead and silver carps. As their fecal matter settles to the bottom, it may be consumed directly by common carp or benthic invertebrates that serve as another food resource for common carp. Foraging common carp mix the benthic sediments and resuspend particulates and increase nutrient availability in the water column, increasing productivity of the planktonic food base for the filter-feeding carps. As the sediments become enriched with nutrients, they will support growth of additional rooted vegetation, providing forage for grass carp, whose feces stimulates benthic productivity in the same manner as the other carps in the pond. By strategically combining different species at the proper densities, wastes can be captured and put to productive use and harvests can be increased for a given level of inputs and water use (Bosma and Verdegem, 2011; Woynarovich et al., 2010). Integrated multitrophic aquaculture (IMTA) expands the polyculture concept to larger food webs and other culture systems, allowing the wastes from higher trophic level species (finfish) to trickle down to lower trophic levels (crustaceans, bivalves, algae) (Soto, 2009). By following a bastardized axiom – "One fish's loss is another fish's lunch" – polyculture and IMTA can capture waste and create value throughout the trophic cascade. Linking aquaculture with traditional terrestrial agriculture, such as culturing fish or shrimp in flooded rice fields, has also proven quite effective in increasing the productivity of both systems (Halwart and Gupta, 2004).

Most of this chapter has focused on freshwater and competition between aquaculture, terrestrial agriculture, and other uses of this precious resource. Nations, decision-makers, factions of end-users, economists, sustainability experts, and others are mired in generations-long disagreements over the best use of freshwater resources, but there are few competing interests that would lay claim to the vast potential of the world's oceans. While mankind debates and occasionally wars over the fate of less than 1% of the water on

Earth, the saline 97% lies comparatively untapped. Just over a third of the world's farmed seafood – roughly 27 million metric tons of fish and shellfish per year – comes from marine and coastal waters (Food and Agriculture Organization, 2016b). Even considering the additional 27 million metric tons of seaweeds produced annually, marine aquaculture is grossly underdeveloped in comparison with inland aquaculture. The flat trajectory of the world's fisheries has led many to conclude that we will not be getting any more food from our oceans; if marine aquaculture is properly developed, those predictions might be happily proven false.

Chapter 11

Aquatic habitat and siting of aquaculture facilities

"The tropical farmed shrimp you eat destroys mangroves." So says the lead-in to a video on the website for the Mangrove Action Project, framed by images of black and yellow backhoes threatening pristine coastal greenery with their gap-toothed buckets (Mangrove Action Project, n.d.). To the swelling soundtrack, images of destruction move across the screen, emblazoned with the message that mangroves are being leveled to make way for shrimp "injected with antibiotics" and growing in "pesticide filled water." The cartoon syringe of purple-colored antibiotics and skull-and-bones pesticide molecules bobbing in the water are silly and they are wrong; perhaps the video's creators intended them as hyperbole or a bit of artistic license or perhaps they used them a bit more cynically. These inaccuracies aside, is the rest of their message off-base? Is farmed shrimp to blame for the loss of mangroves?

Mangroves are halophytes, saltwater specialists that have adapted to withstand the saline water, heat, and wave action of tropical coasts around the world. Their scrubby branches are held aloft by leggy trunks that are continuously inundated by the tides. The submerged structures bear the brunt of tidal action, holding sediments in place and creating a safe haven for young fish and shellfish that would otherwise be tossed and dashed by the waves. Other species seek refuge from predators among the tangle[1] of roots and trunks, but

---

[1] Or perhaps "mangle" would be a better word choice – *Rhizophora mangle* is the scientific name of red mangrove, the most widespread of these salt-loving trees.

predators are attracted nonetheless by the abundance of prey hiding among the mangroves. Mangroves have equal ecological importance above the waterline, providing critical habitat and feeding opportunities for multitudes of coastal birds and mammals (Calow, 1998). Estimates vary, but the current distribution of mangroves is thought to be roughly one-half of its historical coverage, with most of the loss occurring in the last 50 years (Ashton, 2008).

Shrimp farming did not target mangrove forests with any sort of black-hearted intent to disrupt these tropical nurseries and biodiversity hotspots. In fact, the industry did not really target them at all: shrimp farming began in and around mangrove forests simply because that is where the shrimp could be found. The mangroves provided a steady supply of juvenile shrimp that, once fattened in adjacent brackish water ponds, were an easily produced and harvested crop. More recently, shrimp culture has become one of the most lucrative sectors of aquaculture and production has exploded: the world supply of farmed shrimp and prawns has doubled time and again, from only 9000 metric tons in 1970 to more than 4.8 million metric tons in 2015 (Food and Agriculture Organization, 2017c). Construction of brackish water ponds for shrimp culture increased in step with expansion of the industry, resulting in the destruction of mangrove habitat. However, farmed shrimp are not the greatest threat to mangroves: harvest of mangroves for timber and firewood and discharge of industrial and municipal waste also threaten mangroves, but human population growth and the development of coastal wetlands has likely done more damage to mangroves globally than any form of aquaculture (Ashton, 2008). Thailand – one of the world's largest producers of farmed shrimp and widely considered a "poster child" for mangrove destruction – lost one-half of its mangrove forests from 1960 to 1996 (~200,000 hectares); only one-third of the deforested area was developed for shrimp aquaculture. That said, shrimp aquaculture was the primary cause of mangrove deforestation in some other locales, such as parts of Bangladesh, Vietnam, and India (Berlanga-Robles et al., 2011). A more recent assessment of Bangladesh, Brazil, China, Ecuador, India, Indonesia, Thailand, and Vietnam revealed that aquaculture is responsible for about 28% of mangrove loss overall, but with considerable variation among nations. For instance, Vietnam and Thailand have managed to hold on to about 30–40% of their historic mangrove habitat; about one-half of what has been lost is now being used to farm shrimp; in China, less than 10% of historic mangrove stands remain, but more of this has been lost to terrestrial agriculture and urban sprawl than aquaculture (Hamilton, 2013).

Perhaps the common driver of mangrove deforestation is not shrimp farming, forestry, or property development, but rather the failure to recognize the importance of these unique ecosystems in the first place (McDonough et al., 2014). These dense, salty wetlands form a barrier against tidal surge and support fisheries and tourism – they have value beyond their potential to become a shrimp pond or a shopping mall. Unfortunately for the victims of the 2004 Christmas Tsunami, this realization came too late. An undersea earthquake sent waves more than 30 m high hurtling toward the coastal and island nations of Southeast Asia, killing more than 225,000 people and displacing more than 1 million more, with India, Indonesia, the Maldives, Sri Lanka, and Thailand suffering the greatest losses. Beyond these human losses, these countries saw their fishing fleets and aquaculture farms decimated and hundreds of thousands of their residents falling below the poverty line as a result. Where mangroves would have otherwise absorbed some of the tsunami's energy, denuded coastlines suffered the most damage (Forbes and Broadhead, 2007). But hope sprung from the aftermath of the worst natural disaster of modern times: with their coasts scrubbed clean, these countries had an opportunity to start again and correct the mistakes of the past. With international support, affected countries banded together to rehabilitate protective reefs and coastal habitat and rebuild their fishing and aquaculture industries with an eye to sustainability (FAO Regional Office for Asia and the Pacific, 2007). Today, shrimp ponds are more likely to be constructed deeper, with wider embankments, and further inland behind a protective barrier of mangroves, and hatcheries have been rebuilt to provide for greater biosecurity and better quality seedstock (Padiyar et al., 2006).

Shrimp aquaculture has learned the hard lessons of habitat destruction, but is aquaculture to blame for the loss of aquatic habitat in a broader sense? The wholesale transformation of ecosystems that we associate with shrimp farming practices of the past is not typical of aquaculture, but less overt effects on the environment do occur. The most common environmental impacts are eutrophication, the loss of biodiversity or ecosystem services, and the introduction of non-native species or diseases. Each of these topics is addressed elsewhere in this title (see Chapters 10, 12, 13, 16, and 19 for discussions of the inputs and outputs of aquaculture). In addition to these negative effects, it is important to recognize the potential for positive effects of aquaculture on aquatic habitat. Filter-feeding bivalves are often lauded for improving water quality by harvesting untold volumes of plankton and other particulates from the water column, but mollusk aquaculture also improves habitat in other ways (Gallardi, 2014; McKindsey et al., 2006). Suspended stringers and bags of oysters and mussels act as buffers to wave action and add a vertical element to otherwise two-dimensional habitats. They also provide food to the surrounding assemblage of fish and invertebrates, in the form of the gametes and larvae they set adrift and consolidated feces sent sinking to the benthos. Whether the effects are considered positive or negative, the answer to questions of aquaculture-related habitat modification and environmental change may come down to three things: location, location, and location.

Carrying capacity is a fundamental characteristic of ecosystems, describing the number of plants, animals, and other organisms that a given habitat can support in perpetuity without environmental degradation. Carrying capacity is a quantitative metric: a measure of population size or biomass the environment can support. It is also a qualitative metric: a measure of balance, of harmony. Aquaculture is constrained by the ecological carrying capacity of the environment, as well as the social carrying capacity of our communities (Aguilar-Manjarrez et al., 2017). The number of oyster beds that will flourish in a coastal bay is determined as much by the productivity of plankton blooms as it is by the opinions of local landowners and other waterfront entrepreneurs.

Although governments may be motivated primarily by the desire to minimize conflicts with other users of water and space and to limit effects of aquaculture on the surrounding environment, the creation of aquaculture management areas can benefit the aquaculture industry, too (Aguilar-Manjarrez et al., 2017). Zoning may lead to some biologically suitable waters being off-limits to would-be fish farmers because raising fish or shellfish in the space is too controversial. This is a short-term loss for farmers that had their eyes on such spaces, but they are saved the long-term consequences of fighting what is most likely a losing battle to secure and maintain the social license to use these waters. The industry also benefits from avoiding heated public battles between local community members, environmentalist groups, and those intrepid farmers that choose to site their facilities in multi-use or contested waters. Spatial planning can identify sites that are less objectionable to other stakeholders and less risky for farmers. For example, areas that are prone to flooding, droughts, or other severe weather events or at-risk for harmful algal blooms, hypoxia, or pollution can be identified and avoided. Farmers proposing to site their facilities in designated aquaculture-friendly zones may also find it easier to secure financing, as lenders will have greater confidence in the success of operations located in vetted environments.

Increasingly, spatial planning analysis is being used to identify locations that would be biologically suited to aquaculture (Canzi et al., 2017; Huber et al., 2016; Kapetsky et al., 1990). Such tools will help nations move past the question of whether to engage in aquaculture to the more important question, "Where?"

Chapter 12

Escapement from aquaculture facilities and interactions with wild fish

In late August 2017, Washington Department of Fish and Wildlife (WDFW) officials announced a new fishery: all you needed was a valid fishing license, there were no restrictions on fishing gear, the season was open until further notice, and there was no limit on the size or number of fish you could catch and keep. Under most circumstances, Washingtonian anglers would be thrilled at the chance to wet a line and stock their freezers with unlimited fish. A few anglers were probably excited, but WDFW's announcement was not considered good news. The reason? The agency had declared open season on Atlantic salmon because a net pen holding roughly 300,000 fish had just collapsed in Northern Puget Sound (Mapes and Bernton, 2017).

Escapement is primarily a concern for production systems situated in or near open waterways. Net pens and cages are the most vulnerable to breach and escapement. Net pens and cages are usually sited where they will be regularly flushed by strong currents or tides. The moving water removes waste and keeps oxygen levels high in the pen, but also puts the structure under considerable physical strain, particularly during extreme weather events. Predators seeking an easy meal will often attempt to bite or tear holes in the mesh that forms the pen, trying to get in. Less docile cultured species or underfed fish (Glaropoulos et al., 2012) may do the same, trying to get out. Even if they are unsuccessful, they can create weak points that cause the mesh to fail later under physical stress. Coastal areas can also be busy, crowded places, and aquafarms can be unintentionally[1] damaged by commercial or recreational boat traffic or other activity on the water. Escapement may also occur indirectly via eggs or fry produced by reproduction occurring within the culture system (Jensen et al., 2010). Other production systems are less vulnerable, but not impervious to escapement: flooding or catastrophic infrastructure failures can allow fish or shellfish to escape ponds or other land-based aquaculture systems.

Once fish have escaped from an aquaculture facility, what do they do? What harm can they cause? The answers to these questions largely depend on the species involved and whether the same or similar species exist in the environment the fugitives now call home. If cultured Atlantic salmon escape from a damaged net pen into the waters off the Eastern Canadian or American coastlines, it is likely that those fish will encounter wild Atlantic salmon. In fact, this has happened a number of times. Although the cultured fish and wild fish are the same species, they can be said to have divergent biologies (Gross, 1998).[2] Although salmon populations throughout the world were declining many years before the "blue revolution" and subsequent growth of the aquaculture industry, escapement of cultured fish and interactions between wild and farmed salmon have nonetheless been implicated in the continued decline of wild stocks (Ford and Myers, 2008).

Wild salmon are shaped by the invisible hand of natural selection,[3] whereas the genetic makeup of farmed fish is influenced

---

[1] Or intentionally. Sabotage – motivated by anti-aquaculture sentiment, the desire to poach, or for other reasons – has been implicated in a number of escapement losses from net pens (Anon., 2003, 2005; Tukia, 2017).

[2] This author went so far as to suggest – without apparent sarcasm – that a new scientific name, "*Salmo domesticus*", was needed to describe farm-raised Atlantic salmon. This suggestion does not appear to have gained much traction among taxonomists.

[3] Or perhaps "*mostly* natural selection" would be better phrasing. Wild salmon and other organisms are subject to normal evolutionary processes, but must also contend with the pressures of human activity and anthropogenic-related change in the environment. The genetics of a population will come to reflect those individuals that are the strongest and fittest in the brave, new world of fishing, habitat loss, pollution, and other environmental disturbances, but this is not precisely what most consider to be *natural* selection.

by the adaptation to the constraints and rigors of the culture environment as well as selective breeding to favor the largest, fastest growing, most vigorous individuals. What makes a fish successful in captivity – docility despite crowded conditions, acceptance of an artificial environment and prepared diets, neutrality in the face of human or other potential threats, uniformity in pedigree and performance, and so on – is unlikely to translate well to life in the wild. There is a burgeoning body of evidence documenting the molecular, morphological, and behavioral differences between hatchery-origin and wild fish,[4] and the differences between wild fish and those raised for the dinner table are even more stark. If the escapees are fertile and live long enough to interbreed with wild fish, the genetic elements that drive adaptation to the culture environment can find their way into the naturally reproducing population, potentially reducing its genetic diversity and fitness (Baskett et al., 2013). But, there is a bit of a "catch-22" here. If farmed fish are as biologically incompetent as some would argue, it does not seem likely that they would live long enough to pass their genes to subsequent generations. On the other hand, if the fugitive fish manage to reach reproductive maturity and are savvy enough to seek the spawning grounds and attract a mate, perhaps their genes are not such a serious threat to the fitness and survival of their wild brethren. That said, maintaining fitness is as much about retaining rare alleles to support future evolutionary adaptation as it is about maintaining the ability to survive and reproduce under current conditions.

Of course, if the cultured animals are sterile, there is no possibility of polluting the locally adapted gene pool. Many authorities require

---

[4] Describing the differences between hatchery and wild fish has become something of a cottage industry. Understanding the fundamental differences between hatchery and wild fish has driven "hatchery reform" initiatives and greatly improved the ability of natural resource agencies to produce hatchery-origin fish that are better suited to life in the wild and more likely to fulfill management and conservation efforts (Trushenski et al., 2015). That said, the simplicity of the recipe – (1) get some hatchery and wild fish; (2) compare them until you find something, anything statistically different; and (3) publish – seems to have created a bloated and not uniformly insightful body of literature.

that all fish cultured in net pens or other open systems be triploid,[5] rendering them functionally sterile and safe-guarding against most of the reproductive risks of escapement. Of course, there are other ways in which an escaped farmed fish can affect wild fish. Closer proximity because of escapement could theoretically increase the risk of disease transmission to wild fish (Arechavala-Lopez et al., 2013). Given the free movement of pathogens between the ecosystem and open culture systems, though, the disease risk posed by escaped fish is probably not appreciably different from whatever risk they represented when they were still in the net pen or cage (see Chapter 13). A more serious consequence is escaped fish competing with wild fish for food and space.

To simulate the effects of a catastrophic net pen failure and escapement event, groups of farmed Atlantic cod were released into a Norwegian fjord also home to wild populations of cod. Although the farmed fish were observed to rapidly disperse across the available habitat and congregate with wild fish during feeding and spawning, the farmed fish were probably not very effective competitors and were possibly more vulnerable to capture and recovery efforts than their wild counterparts (Uglem et al., 2008). Atlantic salmon escapees appear to readily disperse as well, but other species appear to loiter in and around the farm from which they escaped, suggesting there are some species-specific differences in post-escape

---

[5] Most organisms are diploid (2n), meaning they have two sets of chromosomes: one from mom and one from dad. As readers will undoubtedly remember from high school biology, mom and dad produce haploid (1n) sperm and eggs through a series of reductive cellular divisions called meiosis. For those who remember a bit more from biology class, meiosis II and elimination of the second polar body does not happen in most fish eggs until sometime after fertilization. If a pressure, temperature, or other shock is applied just so, fertilized eggs can be forced to retain the second copy of mom's genetic information that is normally discarded. The result is a triploid (3n) embryo with three sets of chromosomes: one paternal and two maternal. Most of the resultant fish's biology is unaffected, but the presence of a third chromosomal set creates an imbalance that prevents normal meiosis, meaning the triploid fish is effectively incapable of producing viable gametes. They may engage in spawning behavior, but the eggs are duds and the sperm are blanks, meaning their genetics die with them.

behavior (Arechavala-Lopez et al., 2011). Other experimental releases of farmed rainbow trout suggest that some fish may disperse but these tend to die relatively quickly, whereas surviving fish from the same escapement event are those that stay in the immediate vicinity and remain reliant on food resources made available by the farm (Blanchfield et al., 2009). Fortunately, even species like Atlantic salmon that disperse after an escapement event can be captured quickly and effectively: fishermen caught 79% of released Atlantic salmon within a month of an experimental escapement event (Chittenden et al., 2011). Despite appearances, large escapement events are actually easier to manage and mitigate than low-level "leakage" from farms: whereas large escapements are obvious and trigger recapture efforts, low-level escapement may not even be noticed. Somewhat counterintuitively, smaller escapements may also lead to more significant effects on the fitness of wild populations (Baskett et al., 2013).

Organisms escaping from aquaculture facilities can become established in the surrounding environs. If the species is not native to the area, resource managers may have a significantly greater problem on their hands than a just few thousand immigrants mingling with the resident populations. Depending on their biology and the state of the receiving waters, introduced species may go largely unnoticed … or they may escape our control and wreak havoc on a landscape-scale. For instance, sterile grass carp are stocked throughout much of the United States to control burgeoning stands of aquatic vegetation. These voracious herbivores are used to crop nuisance aquatic plants and algae in the same way that goats are used to graze noxious weeds on the land. Unable to reproduce, the sterile grass carp eventually grow old and die or are taken by predators, departing this world without leaving offspring to succeed them. Elsewhere, releases of reproductively intact grass carp have proven more damaging. The same can be said of common carp, stocked widely throughout the United States before fisheries scientists understood the consequences of playing "Johnny Appleseed" with aquatic species.[6]

---

[6] Of course, grass carp and common carp are no longer the most famous invasive carps in the United States: that dubious title belongs to silver carp and bighead

If left unchecked, invasive species can eat or outcompete native species and upend the delicate balances that allow ecosystems to function, causing ecological and economic harm. For example, as aquaculture grew in prominence in Chile, escaped nonnative salmonids became more common. After fugitive salmonids became commonplace in Chilean lakes, native fish populations struggled to persist under the less-favorable ecosystem dynamics established following the introduction of these new predators (Arismendi et al., 2009). Conversely, escapement events and intentional introductions of farmed tilapia and carps throughout Southeastern Asia appear to have had little effect on native fish assemblages (Arthur et al., 2010). Estimates vary considerably because of the complexities of invasive species accounting (Lovell and Stone, 2005), but one analysis suggests that the 138 nonnative fish introduced to the United States are responsible for endangering 44 native species and causing US$1 billion in economic losses *annually* (Pimental et al., 2001).[7] Economists and ecologists may debate a few million dollars or the odd imperiled species here or there, but the consequences of aquatic invasive species on native ecosystems and the economies they support are significant, and in many cases, severe. Once an invader becomes established, it is next to impossible to eradicate them – the best resource managers can hope for is to control the population, limit the damage, and stop the next introduction before it happens.

There are a number of pathways for aquatic species introductions, including aquaculture. As it happens, aquaculture is part of how all of the nonnative carps came to the United States. A few Americans had previously kept and occasionally released common carp to nearby waters, but the species did not become established in the United States until the U.S. Fish Commission began producing

---

carp. Fecund, large-bodied, and with a tendency to take to the air, the bigheaded carps have become a dominant feature in the Mississippi River and its tributaries – a swarm of invaders pushing upstream in search of food and less crowded waters.

[7] This staggering figure is a net estimate: the economic gains provided by some invasive species (those that support fisheries, that is) are taken into account, but the balance still is overwhelmingly negative.

them *en masse* in the 1870s and 1880s and stocking them in every river, stream, lake, and large-ish puddle (Nico et al., 2014).[8] Nearly a century later, resource managers were more circumspect about introducing fish to American waters, but only just: grass carp were imported by the U.S. Fish and Wildlife Service and widely propagated and stocked by federal and state agencies throughout the 1960s and 1970s for aquatic vegetation control.[9] Silver and bighead carps were imported in the early 1970s to control phytoplankton and zooplankton blooms in aquaculture ponds and wastewater treatment lagoons in the Southeastern United States. These fish were never intentionally stocked in open waters, but flooding allowed them to escape into the Mississippi River drainage (Kolar et al., 2005). It is important to distinguish between the stocking of hatchery-produced common and grass carps and the unintended escape of bigheaded carps from aquaculture ponds. The former were ill-advised, but intentional actions of state and federal natural resource agencies; the latter was an accident caused by floodwaters that breached pond levees. There may be little value in parsing culpability, but there is a difference between allowing a fish to escape a private pond and willfully introducing it to public waters.

The failed net pen in Puget Sound in August 2017 was discovered before most of the fish could escape, but as many as 100,000 were released. Anglers and tribal fishermen captured a good number of fish in the following weeks. As time passed, fewer fish were caught

---

[8] At one point, the U.S. Fish Commission was distributing common carp across the landscape by the railcar load, literally. Specially designed railcars were fitted with compartments to hold containers full of juvenile fish and ice to keep fish and fish eggs cool during warm weather. Whenever the train crossed what looked to be a suitable body of water, carp were released in the hope that they would be prosperous and grow to feed families of European descent settling across the American countryside (Nico et al., 2014).

[9] The first grass carp were imported to Stuttgart, Arkansas, in 1963. Accidental releases were documented as early at 1966, but these were not considered especially serious at the time, as fertile grass carp were being stocked in open waters nationwide throughout the 1960s and 1970s. It was not until the first successful triploid fish were produced in 1983 that agencies began to use sterile fish for control of aquatic vegetation (Mitchell and Kelly, 2006).

and those that were had empty stomachs. Unable to successfully transition to a natural diet and evade predators in the Sound, by the end of 2017 the majority of fugitive salmon were presumed dead (Mapes, 2017). Biologically, the consequences of the escapement were minimal, but the economic and political consequences for the net pen owner and other operators in the Sound depend on the outcome of the legal fight that is brewing (Camden, 2018). Atlantic salmon seem particularly ill-suited to a fugitive life "on the lam" in Puget Sound, but this is not much of a consolation to those who want to see the aquaculture industry succeed and win over the hearts and minds of the public. The next escapement – in Puget Sound or elsewhere – might not be so ecologically innocuous. Best for all concerned to keep the fish in the kettle.

Chapter 13

Disease transmission between farmed and wild fish

Infectious agents are nothing new in the ocean; pick up any marine species and you can find a parasite or a pathogen, or perhaps a dozen with no cause for alarm. (Lafferty and Hofmann, 2016: 1)

Perhaps the first aspect of a normative or normal condition for pathogens might be considered their ubiquity. (Coutant, 1998: 102)

In February 2012, a news item about fish disease appeared in *Fisheries* magazine, the flagship publication of the American Fisheries Society. Reprinted from an external source, the short piece described the discovery of a fish virus, infectious salmon anemia virus (ISAv) in the coastal waters of British Columbia in late 2011 and linked the pathogen's presence to salmon farms in the area. In collaboration with Alexandra Morton, a prominent, if scantly credentialed anti-aquaculture activist,[1] researchers at Simon Fraser University claimed to have found the lethal virus in two feral sockeye salmon juveniles. The piece also alleged the U.S. National Marine Fisheries Service and Fisheries and Oceans Canada were attempting to suppress these findings. A deadly virus, leaked

---

[1] She is occasionally referred to as "Dr. Morton", but Ms. Morton's doctorate is an honorary degree bestowed upon her in 2010 by Simon Fraser University for her wild salmon activism (Simon Fraser University, 2010). Sometimes described as a "self-taught" biologist, the only non-*honoris causa* degree Ms. Morton holds is a bachelor's degree from American University.

documents, political intrigue, a cover-up – the piece had all the trappings of an international scandal. There was just one problem: it was not true. The journalistic rigor that led to reprinting of this piece in *Fisheries* was dubious at best, and the staffer responsible was swiftly dismissed following retraction of the article. The story did not end with the editorial staff's red-faced admission of their error: just over two and a half years later, *Fisheries* magazine published another article about ISAv, this time a synopsis of the joint response of American and Canadian authorities to what had been zealously, but mistakenly reported as the first discovery of ISAv in the Pacific Northwest.[2]

The team claiming to have detected ISAv in British Columbia used a molecular technique called real-time or quantitative polymerase chain reaction (qPCR). Although polymerase chain reaction (PCR)-based techniques are useful as a confirmatory test, often used to differentiate between closely related species or strains of microbes, they are rarely used to screen for the presence of pathogens. This is because PCR-based techniques, unless coupled with other more discriminating assays, are exceedingly prone to false-positives. PCR methods work a bit like biochemical copy machines, capable of creating millions of copies from a single original fragment of DNA. Once enough of the genetic material has been generated by the PCR reaction, it can be compared to known samples of DNA to determine if the original sample was a match. Although the technique was once considered unfamiliar and potentially unreliable,[3] over the

---

[2] ISAv is principally a disease of Atlantic salmon, though ISAv infections have been experimentally induced in other species. In affected fish, ISAv causes severe anemia, organ failure, and ultimately death. ISAv has caused widespread losses in farmed Atlantic salmon in Norway and Chile and has been reported in the United Kingdom, the Faroe Islands, Eastern Canada and the United States (New Brunswick and Maine, respectively), but wild Atlantic salmon, brown trout, and sea trout (anadromous brown trout) are thought to be important carriers of the disease (World Organization for Animal Health, 2016).

[3] For many, their first introduction to PCR techniques and genetic fingerprinting was the prosecution of O.J. Simpson for the murders of Nicole Brown Simpson and Ronald Goldman. Although the scientific community considered the validity of genetic fingerprinting to be settled science (Gladwell, 1995), public opinion

past 20 years genetic "fingerprinting" has become an accepted technique and is widely used in forensic science, biomedicine, and many other fields.[4]

PCR techniques are amazingly powerful, but their sensitivity is also their weakness: PCR tests are sensitive enough to pick up even a single strand of DNA, but they cannot tell you whether the genetic material is fresh or decades old. They also cannot tell you whether the material came from a living cell or organism: dead or alive, DNA is all you need for a positive PCR result. Matching a suspect's DNA to a smear of blood from a crime scene with PCR-based DNA fingerprinting is powerful evidence of their involvement in the crime. But what if there were no smear of blood? Unless the crime scene had been recently and thoroughly cleaned with bleach, it is likely that swabs of the crime scene would yield a great number of genetic fingerprints.[5] Without corroborating evidence suggesting the suspect was involved – a relationship with the victim, an eyewitness account, or blood or other trace evidence found at the scene – the positive PCR result is not very convincing.

---

was decidedly more ambivalent. Despite the strength of the DNA evidence linking Simpson to the scene of the murders, jurors were ultimately unconvinced. Some 20 years later, the genetic experts on Simpson's legal team, Barry Scheck and Peter Neufeld, are perhaps now better known (some would say, ironically) for having founded the Innocence Project, a non-profit organization dedicated to the use of DNA testing to exonerate those wrongly convicted of serious crimes (Hardy, 2009).

[4] Harnessing PCR truly revolutionized basic and applied genetic research. PCR-based techniques have since come out of the laboratory and are modern television mainstays, featured in the crime labs of virtually every prime-time police procedural and the paternity tests of daytime talk shows. Today, services such as 23andMe and Ancestry DNA let consumers do their own forensic science and use PCR-based techniques to learn about their ethnic heritage and find distant relatives.

[5] Here the term "fingerprint" has two meanings: taken to the extreme, PCR-based methods can detect so-called touch DNA, transferred from skin to surfaces by the briefest of contact. Given the potential for erroneous or inconclusive results and misinterpretation of positive results, touch DNA tests are controversial and are not used routinely in criminal prosecutions in the manner of DNA evidence derived from blood or other bodily fluids (Inman and Beck, 2011).

What the team at Simon Fraser University found in two sockeye salmon was analogous to "touch DNA" collected in the absence of other corroborating evidence. ISAv is, first and foremost, an Atlantic salmon disease – sockeye salmon are not a typical host for the virus (World Organization for Animal Health, 2016) and Pacific salmon are generally considered less susceptible to ISAv (Amos et al., 2014). What is more, the two juvenile fish that were reportedly positive for ISAv were not showing any clinical signs of infection. The sockeye salmon tested were part of a larger study in which the researchers found "non-negative" results in less than 2% of the fish sampled, and these were not an exact match to known genetic fingerprints of ISAv. Finally, none of the isolates from reportedly PCR-positive fish yielded positive results after exhaustive attempts to detect ISAv using the traditional method of cell culture – a means of determining whether any live viruses are present. As the researchers acknowledged in the article that was eventually published, "All virus isolation attempts on the samples were negative, and thus the samples were considered negative in terms of the threshold trigger set for Canadian federal regulatory action" (Kibenge et al., 2016: 3). Using the crime scene analogy, there was no relationship between the suspect and the victim, no eyewitness testimony, no corroborating trace evidence, and the DNA fingerprint evidence was questionable – in fact, in the case of ISAv in British Columbia, there was not even much to suggest that a "crime" had been committed in the first place. In the subsequent years, American and Canadian authorities have conducted intensive surveillance for ISAv in Washington, Alaska, and British Columbia and have found no evidence of the virus in Pacific waters.

The myth of ISAv in the Pacific Northwest is, for now, just that – a myth. But the story was reported again and again and widely accepted as fact, in part because it resonates with a firmly and widely held misconception: hatchery fish are sickly creatures whose wickedness threatens unsuspecting, otherwise vigorous wild fish with infectious disease. Even seasoned fisheries professionals fail to grasp or articulate the basics of aquatic disease ecology and pathogen transmission. Dick Vincent, former regional fish manager and whirling disease coordinator with Montana Fish Wildlife and Parks,[6] once remarked of the fish disease that made him famous, "When I first heard of whirling disease, I didn't pay much attention to it," Vincent said. "I thought, 'OK, it's probably a hatchery disease and not really in the wild'" (Boswell, 2009). The inaccuracy of Vincent's statement can be chalked up to an offhand comment made without much thought, but coming from a recognized expert it is shockingly erroneous. Charlie Smith, a veritable legend in the practice of fish health in North America and a contemporary of Vincent's, commented that, "Vincent and some of the others back then were so anti-hatchery, there was hardly any talking to them. There's no such thing as a 'hatchery disease' that doesn't occur in wild fish" (C. Smith, interview with J. Trushenski, 2017). The timing or severity of an infection might be different in a hatchery raceway compared to a stream, but the pathogens – the infectious agents themselves – come from nature, from the ecosystems of which they are a part. Infectious diseases influence the economics of capture fisheries and aquaculture operations alike, by reducing the biological productivity of the aquatic organisms themselves and/or the commercial value of the seafood they yield. In the absence of biosecurity protocols to limit host/pathogen interactions, these diseases are readily transmitted from farmed fish to wild fish *and* from wild fish to farmed fish (Lafferty et al., 2015).

Whirling disease, named for the effect it has on infected fish and their swimming behavior, is a significant disease, in terms of biology and policy. Following the 1996 discovery of whirling disease in the intermountain West, the U.S. Fish and Wildlife Service was charged with assessing the health status of wild fish populations. Fish diseases had been studied in the wild before, but never in a comprehensive manner, and information on the distribution and prevalence of fish disease in the United States was scant. For more than 20 years, the U.S. Fish and Wildlife Service has operated the National Wild Fish Health Survey. In cooperation with various

---

[6] Over the course of his career, Vincent was a staunch critic of fish hatcheries and stocking programs. He is widely credited with Montana's 1974 decision to stop stocking hatchery fish where wild trout populations existed (Vincent, 2004).

partners, including states, tribes, and the private sector, well over 3 million fish have been evaluated for just short of two dozen bacterial, viral, and parasitic pathogens (U.S. Fish and Wildlife Service, 2017). Although the survey originally focused on a dozen groups of fish, mostly of recreational or commercial importance, as of April 2017, 275 unique fish taxa have been surveyed across 49 states and the District of Columbia.[7] Pathogens were positively detected in 2878 of the 167,245 unique cases processed to date, or about 1.75%. These results would suggest a very low prevalence rate for all of these pathogens, but it is important to recognize the limitations of the dataset, specifically the number of fish species and pathogens it represents and the fact that most of the sampling was done opportunistically at the researchers' convenience and involved fish that were, to all outward appearances, healthy. The contiguous United States is currently home to more than 3000 species of finfish (Froese and Pauly, 2017) and standard methods exist for routine detection of more than 17 bacterial, 14 viral, 18 parasitic, and 2 fungal pathogens of concern throughout North America and countless other potentially pathogenic microorganisms (Fish Health Section of the American Fisheries Society, 2016).

Why does the public associate disease with aquaculture, but not wild fish? Øivind Bergh (2007) posed this question in an essay entitled, "The Dual Myths of the Healthy Wild Fish and the Unhealthy Farmed Fish." Part of the answer is that it is inherently difficult to study disease in wild fish. Infected fish are compromised and often more vulnerable to predation.[8] Whether they are caught and consumed by predators or succumb to infection, diseased fish often die before they can be observed or captured for study. It is infinitely

---

[7] The database does not include records for fish collected in Hawaii.

[8] Parasites that must infect multiple hosts to complete their life cycle are especially ruthless – or perhaps ingenious – in this respect. Parasitic flukes, for example, can cause erratic swimming and other conspicuous behavior in intermediate fish hosts, making them more vulnerable to predation by piscivorous birds, the fluke's definitive host. Rather than leaving things to chance, the fluke increases its odds of infecting an avian host by making its current host more obvious to would-be predators. The phenomenon is sufficiently common to have its own terminology: parasite increased trophic transmission (Lafferty, 1999).

easier to observe and sample fish for diseases in the captive environment of a hatchery or a fish farm. Much of what is currently known about fish diseases comes from work done in the aquaculture setting, but this does not mean that diseases are absent in the wild. Hatcheries can create conditions that favor disease, but they are hardly the source of them, says Doug Burton, a fish pathologist recently retired from the Idaho Department of Fish and Game (IDFG). "Hatchery fish are reared in an artificial environment, unnaturally crowded together, fed a diet that only vaguely resembles what they would eat in nature, and are routinely physically handled in ways that would never happen to wild fish. The total of all these 'stresses' can compromise the natural defenses a fish may use to fight disease, and you have the perfect opportunity for an outbreak" (D. Burton, interview with J. Trushenski, 2017). Quoting fellow IDFG pathologist, Doug Munson,[9] Burton adds, "Spontaneous generation was disproven a long time ago. No pathogenic organism ever evolved from concrete" (D. Burton, interview with J. Trushenski, 2017).

Disease is a part of life, for all organisms. Most people simply do not notice fish diseases unless they cause massive die-offs, and although mass mortality events undoubtedly occur in the wild, they are somewhat more likely to occur in farmed fish than in feral populations. Intensive aquaculture lends itself to disease primarily by creating artificially large and dense congregations of aquatic organisms. The density of potential fish or shellfish hosts in aquaculture benefits pathogens: whether they are in a net pen or a preschool, infections can take hold and spread more rapidly when there is an abundance of prospective hosts in close proximity. In the wild, a diseased fish may die before it comes into close enough contact to infect another fish; this is considerably less likely in intensive aquaculture, even under the best of conditions. Should the captive population become immunocompromised by declining environmental conditions, improper husbandry, necessary handling or

---

[9] The author had the privilege of working with the three IDFG fish pathologists from 2015–2017 – the third is sadly not also named Doug, but David Burbank. All three are certified as Aquatic Animal Health Inspectors by the Fish Health Section of American Fisheries Society, of which Munson is also a past-president.

transportation, or other stressors, conditions become even more favorable for pathogens. Hatcheries and farms can serve as reservoirs for infectious agents, but there is little to suggest diseases are frequently transmitted from captive populations of aquatic organisms to feral ones (LaPatra and Foott, 2006). The opposite, however, is observed with some regularity. For example, Chinook salmon carrying *Renibacterium salmoninarum* that spawn naturally in the streams that supply hatcheries with water are known to contribute to outbreaks of bacterial kidney disease in the hatchery-reared fish (D. Munson, interview with J. Trushenski, 2016). Wild and farmed fish are known or suspected to exchange a great number of infectious agents, but examples of pathogens that are only passed in one direction – exclusively from farmed fish to wild fish or the other direction – are all but unknown. When exotic pathogens are introduced to naïve populations as a result of aquaculture activities, the consequences are often clearly negative; the effects of aquaculture on endemic pathogen and host communities are anything but (Lafferty et al., 2015).

Until relatively recently, the risk associated with moving biota from one place to another was not widely recognized. Starlings, one of the most common songbirds in the United States, were intentionally released in New York City in 1890 by a man hoping to introduce all of the birds named in Shakespeare's works to the New World (Love and Stamps, 2008). Apples, indigenous to Central Asia, were brought to North America by European colonists and widely planted across the countryside, most famously by John Chapman (Marzec, 2004).[10] One could dismiss the actions of an ornithologically inclined Bard enthusiast and an altruistically green-thumbed folk hero, but even natural resource experts were once unaware of the delicate balance of ecosystems and the disruptive effect of nonnative species. Today, the U.S. Fish and Wildlife Service spends about US$100 million per year on efforts to control and eradicate invasive species (U.S. Fish and Wildlife Service, 2012), but it was the

---

[10] Mr. Chapman is better known to American schoolchildren as "Johnny Appleseed", a nickname that has since become synonymous with the indiscriminate introduction of plants or animals to new landscapes.

U.S. Fish Commission – forerunner to the U.S. Fish and Wildlife Service – that lead an unprecedented campaign[11] to stock American waters with nonnative common carp in the late 1800s (U.S. Fish and Wildlife Service, 2014) and cooperated with state agencies to expand the native range of rainbow trout to include as much of the country as possible (Halverson, 2010). Introducing species outside of their normal distribution often has unintended consequences, including the possibility of introducing novel pathogens to naïve populations not adapted to cope with the disease.

Gyrodactylus salaris, a parasitic fluke, causes little harm to its natural host, Baltic salmon, a collection of geographically and evolutionarily distinct populations of Atlantic salmon. However, when the parasite was inadvertently translocated from Sweden to Norway with a shipment of fish, it caused significant disease outbreaks in hatcheries and rivers. Hatcheries were, of course, implicated in these events, though evidence suggests that fishermen played an equal or greater role in the movement of fish and *G. salaris* in Norway (Bergh, 2007). Although the story of whirling disease is longer and more convoluted, it is essentially the same tale of exposing fish with few natural defenses to a new pathogen. The parasite *Myxobolus cerebralis*, the causative agent of whirling disease, was identified in Germany in 1903, but was not associated with a disease state until 1983, when it was found in rainbow trout and brook trout introduced to Germany in an attempt to bolster the country's declining brown trout populations. The disease caused catastrophic losses at trout farms across Germany, but only among the introduced trout; native brown trout appeared relatively impervious to the disease. With little understanding of how the parasite was transmitted,

---

[11] Carp were stocked with the intention of creating a food source for a growing population, including large numbers of European immigrants who were familiar with the fish and held it in high esteem. The scale of the U.S. Fish Commission's effort was impressive: carp fingerlings were transported across the country by the railcar-load and fish were released whenever the train crossed a water body. Whether by federal, state, or private entities, records document stocking of common carp in every state except Alaska (U.S. Fish and Wildlife Service, 2014).

biologists continued to move fish from place to place, and over the ensuing decades *M. cerebralis* spread across Europe to Asia and North America and beyond (Bartholomew and Reno, 2002; Blaylock and Bullard, 2014).[12] Over evolutionary time, *M. cerebralis* established a stable, if unwilling partnership with its natural host, brown trout. Only when the comparatively defenseless North American salmonids were introduced to the parasite on its home turf did the disease become apparent; only when *M. cerebralis* was introduced to naïve populations around the world were the implications of whirling disease fully realized.[13] As with *G. salaris*, the story of whirling disease is not so much evidence of aquaculture as the "Trojan horse" of fish disease as it is a cautionary tale about the perils of moving organisms from one place to another without appreciating the biological and ecological risks (Bergh, 2007). "Cultured fish probably take the greatest blame for introduction of pathogens to naïve populations, simply because they are the most commonly moved, in the greatest numbers, and with the most documentation," says Burton. "It's not that it doesn't happen in other ways or that moving cultured fish is even the most common cause, but the other ways pathogens move themselves around are much more difficult to detect or measure, and hence, to blame" (D. Burton, interview with J. Trushenski, 2017).

Armed with modern methods of pathogen detection and a greater appreciation of the consequences of invasive species, aquaculturists and aquatic animal health specialists are now largely able to avoid the mistakes of the past. In the industrialized world, aquaculture operates within a regulatory framework designed to minimize the risk of pathogen introduction and disease (see Chapter

---

[12] The first diagnosis of whirling disease in the United States occurred in 1956, after brook trout at a Pennsylvania fish hatchery were exposed to *M. cerebralis*, most likely via frozen food fish or live brown trout imported from Europe. From Pennsylvania, the parasite traveled with live fish transported across the country, reaching hatcheries and natural waters in Idaho, Oregon, Colorado, and Wyoming in the late 1980s (Bartholomew and Reno, 2002).

[13] Although recruitment was undoubtedly affected shortly after the introduction of *M. cerebralis*, affected North American salmonid populations seem to be developing resistance and establishing a new equilibrium with the parasite (Miller and Vincent, 2008).

19). Routine surveillance testing, biosecurity protocols, vaccination against known pathogens, and other preventative strategies greatly reduce the incidence and severity of disease at aquaculture facilities; rapid disease diagnosis and effective therapeutic treatments help to contain and resolve outbreaks quickly and minimize losses and the risk of infecting wild fish populations (Jones et al., 2015). As noted by Bergh (2007: 163), "Often it is not cultured fish that are the most susceptible [to disease], due to efficient prophylactic strategies and good culture practices." Rigorous fish health management reduces, but does not eliminate the possibility of aquaculture affecting wild fish, as may occur with sea lice *Lepeoptheirus* and *Caligus* spp. infestations. As the name implies, sea lice are a natural parasite of marine salmonids, including Atlantic and Pacific species. Fish become infested after coming into contact with a planktonic form of the parasite. After attaching to the fish's skin, salmon lice mature through a series of life stages while feeding on the fish's mucus coat and bodily fluids, ultimately reaching sexual maturity and producing eggs that hatch into the next generation of planktonic parasites. Usually adult fish are not seriously affected by sea lice, though the small wounds they create during feeding can provide other pathogens such as bacteria with a point of entry. Juvenile salmon – called smolts – are at greater risk because of their smaller size: the physical damage caused by sea lice can limit the ability of smolts to swim and maintain proper electrolyte balance and heavy infestations can be lethal (Boxaspen, 2006).

Sea lice infestations of wild fish have been reported for centuries, well before salmon aquaculture became established anywhere in the world, and contemporary assessments of wild salmon populations indicate that the parasites are abundant and prevalent throughout the Northern hemisphere (Jones et al., 2015); similar parasitic copepods regularly infest feral and farmed salmonids in the Southern hemisphere. Net pen operations in Norway, Scotland, and eastern Canada placed farmed salmon in close proximity to wild fish carrying *Lepeoptheirus salmonis* and outbreaks unsurprisingly occurred among the farmed fish. At the same time that salmon aquaculture was expanding worldwide and sea lice outbreaks were becoming more frequent, many fisheries were declining along with

trust in public institutions and the scientific community to do anything meaningful about it. While pro- and anti-aquaculture factions lobbed bilious attacks back and forth, a significant body of
scientific literature was emerging (Blaylock and Bullard, 2014). The
likelihood and severity of sea lice infestation depends on a number
of environmental factors, such as water temperature, salinity, the
strength of water currents and tidal flushing, and so on, but existing
host and parasite biomass are also critically important variables for
the spread of sea lice (Jones et al., 2015) and other aquatic pathogens (Krkošek, 2017). If farmed salmon become infested and are left
untreated, the farm can become a parasite "hot spot", increasing the
number of parasite larvae in the water. Maintaining high host densities within the farm perpetuates the cycle of infestation, creating
a parasite reservoir and extending the period of time during which
sea lice infestations would normally occur in the wild. Although
wild fish are the original source of the parasite, salmon farms can
amplify the threat of sea lice to wild fish (Torrissen et al., 2013).
Consequently, restrictions are now in place to curb the contributions of salmon farming to sea lice infestations in wild populations.
These include regulations that restrict aquaculture development in
sensitive areas and periodic fallowing of existing aquaculture sites.
Other regulations require routine monitoring of farms for sea lice
and management action[14] when the number of infested fish and
parasite counts rise to an unacceptable level. (Jones et al., 2015).

For sea lice and many other diseases of aquatic organisms, minimizing effects on wild fish mostly boils down to doing what is also
best for the aquaculture operation: keeping farmed fish healthy and
free of disease through comprehensive fish health management,
including strict biosecurity and other preventative measures, regular

---

[14] Typically, this means treating with an oral anti-parasitic medication (similar
to that commonly used in the treatment of lice, mites, or worms in humans or
companion animals) or applying a chemical bath (hydrogen peroxide is commonly
used) to kill the sea lice. Although historically either approach has worked well,
particularly when applied synchronously at adjacent farms, sea lice are able to
develop resistance to these treatments and so their long-term effectiveness remains
a concern.

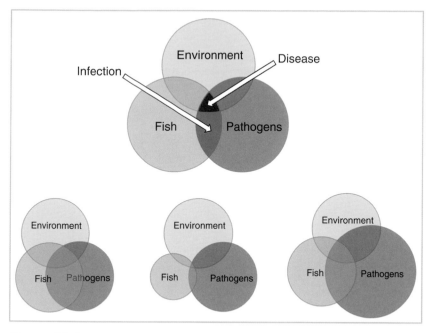

Figure 13.1: Fish health professionals use – with almost comic frequency – a Venn diagram to illustrate the relationships between fish, pathogens, and the environmental conditions both experience. Infection can occur when fish and pathogens overlap, but disease and mortality usually do not unless the environmental conditions conspire to weaken the fish's immunological defenses. Even if one assumes the environment is stable, changes in the number or distribution of fish and/or pathogens influences the likelihood of infections and disease; if the effects of climate change are taken into account, interactions between fish and their pathogens become even more unpredictable (Chiaramonte et al., 2016).

and rigorous surveillance to facilitate early detection, and rapid response in the event of disease. The biggest threat to fish health in the future might be its very uncertainty. "Environmental change is the biggest threat," says Burton. "This may include increasing water temperatures, pollution, increased demand for water resources, or some other factor that is or isn't human-caused. Disease occurs at the center of the classical Venn diagram showing the intersection of fish hosts, their pathogens, and the environment they share [Figure 13.1]. As the three circles change … well, it's anyone's guess what might happen" (D. Burton, interview with J. Trushenski, 2017).

Chapter 14

Feeding fish to fish – use of marine-origin resources in aquaculture feeds

Tell me what you eat, and I will tell you what you are. (Jean Anthelme Brillat-Savarin)

Man is what he eats. (Ludwig Andreas Feuerbach, *Concerning Spritualism and Materialism*, 1863/64)

You are what you eat. (Victor Lindlahr, *You Are What You Eat*, 1942)

When Brillat-Savarin meditated on the influence of man's diet on man himself, his thoughts were of the nature of both: if one wants to be healthy, one must consume foods that are similarly wholesome and nutritious. Subsequent iterations of Brillat-Savarin's famous words, including the now cliché phrase, "You are what you eat," were also meant to invoke less than literal interpretations. For many species of fish, however, the phrase is more than metaphor: these fish, which include a great number of those cultured throughout the world, eat fish. Feeding fish to fish is a practice that harkens back not only to the origins of aquaculture, but to the origins of the species themselves. For piscivores – fish that eat fish – nothing could be more natural than a diet of fish. Feeding fish to fish has become a lightning rod issue among aquaculture's critics, however, and farms using feeds rich with fish-derived protein and fat risk difficulty in achieving sustainability standards and certifications.[1]

---

[1] Use of wild fish in aquaculture feeds, either directly or in the form of rendered fish meal or fish oil, typically counts against sustainability in scoring systems,

How did the natural habits of fish-eating fish, a menu refined over countless millennia of evolution, become an unsustainable choice for aquaculture?

Before examining issues of safety, ethics, and efficiency, we must understand the practice of feeding fish to fish. Low-value fish may be fed directly to the cultured species live, fresh, or after freezing. Fish used as forage (sometimes referred to as "trash fish") are typically small and undesirable as human food. Feeding fish to fish this way is relatively uncommon: sourcing, storing, and distributing fish for this purpose is impractical for most operations. Having to catch or purchase fresh fish for every feeding is inconvenient and costly, in terms of the purchase price or labor needed to do the fishing. Cold storage can alleviate some of the inconvenience, but is itself costly and not always readily available in the remote locations where aquaculture farms are often located. Feeding live or fresh fish is also ill-advised for other reasons largely unrelated to the concerns of sustainability accountants. First, those who rely on wild-caught forage to feed their fish are bound to a resource that can be variable and unpredictable. Most fisheries are seasonal and the nutritional value of wild fish (especially levels of protein and lipid [fats and oils]) is known to vary from month to month and year to year. Second, wild fish can introduce parasites, bacteria, viruses, and other sources of disease to a farm. Freezing trash fish will reduce, but not eliminate pathogens. For most operations, feeding fish to fish this way is too expensive and risky to justify. Trash fish are still used in commercial aquaculture, but typically only in situations where prepared diets are not available or the nutrient requirements of the cultured species are unknown. Forage fish are still used to help transition finicky fish to prepared diets and in some types of capture-based aquaculture operations, whereby wild fish are caught as juveniles and reared to market size in captivity (Lovatelli, 2008). For example, forage fish are still used to fatten wild-caught bluefin tuna, but such "ranching" operations are eager to shift to prepared diets and research to this effect is active and ongoing.

---

such as that used by the Monterrey Bay Aquarium Seafood Watch® program (Monterrey Bay Aquarium, 2015).

The vast majority of forage fish fed by the aquaculture industry is in the form of fish meals and oils. Both products are rendered from forage fish and, increasingly, bycatch and seafood processing scraps and offal.[2] Reduction fisheries – those that harvest fish for rendering – target various species of anchovy, herring, sardine, menhaden, and other bony, oily fish with relatively little value as seafood. The rendering process varies somewhat based on the raw materials and desired end-products involved, but in simple terms, raw fish are cooked, pressed to remove liquids, and dried. The resulting protein-rich powder is fish meal. If the raw fish were sufficiently oleaginous, fish oil can be purified from the liquid fraction pressed from the protein "cake" during rendering. Every year, reduction fishery landings and seafood wastes are rendered to yield approximately 5–6 million metric tons of fish meal and less than 1 million metric tons of fish oil (Food and Agriculture Organization, 2008). Reduction fisheries are some of the most carefully managed marine stocks in the world and landings have been relatively stable for decades, fluctuating only in response to the effects of El Niño Southern Oscillation (ENSO) events on ocean productivity.

Supplies of fish meal and oil have been remarkably consistent from year to year for the past several decades, but demand for these products has skyrocketed, mostly due to growth of the aquaculture industry. Fish meal, or crude versions of it have been used as fertilizer and in animal agriculture for centuries. Indeed, feeding of dried fish to livestock in one Middle Eastern city was sufficiently noteworthy to be mentioned in the "Travels of Marco Polo" at the turn of the 14th century: "… they accustom their cattle, cows, sheep, camels and horses to feed upon dried fish, which being regularly served to them, they eat without any sign of dislike" (Windsor, 1971). Industrialized fish rendering began in the early 1800s as means of handling seasonal surpluses of fish in Europe and North America. Fish oil, then considered the principal product, was used in soap-making, tanning, and other industrial

---

[2] More than one-third of fish meal/oil supplies are derived from fishery residues; by 2022, it is estimated that one-half of fish meal and oil will come from bycatch and processing wastes (Food and Agriculture Organization, 2014).

applications, as well as for production of margarines; the remaining meal was turned into the ground to fertilize crops or, later, fed to livestock (Food and Agriculture Organization, 1986). Marine rendering maintained this oil-centric focus through the first half of the 20th century, but emphasis began to shift to fish meal as it became increasingly important as a feed ingredient for intensively reared terrestrial livestock. Though fish meal is still an important ingredient in weaning and other specialty diets for terrestrial animals (especially piglets), in the 1990s, fish meal producers began to pivot to the growing aquaculture feed market. At the same time, fish oil marketing shifted away from industrial applications to the unique nutritional value of this ingredient to humans and animals. By 1995, the aquaculture industry was consuming a little more than one-quarter of the annual fish meal supply and one-third of fish oil. By 2005, aquaculture's share had increased to nearly 70% of fish meal and 90% of fish oil produced annually throughout the world (Tacon and Metian, 2008).

Aquaculture's growing monopoly of fish meal and oil does not reflect greater and greater incorporation of these ingredients in aquaculture feeds – actually the opposite is true. Twenty years ago, salmon and marine fish, the most significant consumers of marine-derived resources, were fed diets containing roughly 45–50% fish meal and 18–25% fish oil. Contemporary diets for these species contain, on average, 30–32% fish meal and 8–20% fish oil (Tacon and Metian, 2008). The aquaculture industry has come to dominate the fish meal and oil markets for a number of reasons, including industry growth, shifting species demographics, increasing reliance on feed inputs, and the inability of many sectors to easily transition to alternative sources of protein, lipid, and energy. From the mid-1990s to the present day, the aquaculture industry grew roughly threefold: in 1995, global production totaled approximately 25 million metric tons; today, production is approaching 75 million metric tons (Food and Agriculture Organization, 2016b). The industry has grown dramatically, and has also intensified. When the "blue revolution" began, the majority of species were "unfed": carps, shrimp, and others were kept in well-fertilized ponds and subsisted on plankton blooms and other natural foods until they reached the desired size. About a third of the world's farmed fish is still produced this way, particularly in the developing world. But with the addition of supplemental or complete feeds, fish grow faster and yields increase; with direct feeding, ponds that previously provided only enough fish for the family can provide fish for the table and for the market. Common, grass, and black carps are not filter feeders, but the vast majority of farms raising these fish in 1995 did not feed them directly: more than 4 million metric tons – 80% of production – were produced without feeds. By 2010, one-half of these carp were being fed directly and feed use in this sector had grown fivefold (Tacon et al., 2011). Some types of fish, like salmon, trout, and seabass, have always been raised as "fed" species. It is not a question of whether to offer feed to these fish – feed is a requirement for their successful culture. As the popularity of fed species has increased and growers have increased their use of feeds, even for traditionally unfed species, demand for aquafeeds and their constituent ingredients has grown even faster than the aquaculture industry itself. In 1995, a total of just over 4 million metric tons of fish and crustaceans were raised, using approximately 7.6 million metric tons of feed; by 2008, over 29 million metric tons of feed were being used to raise nearly 17.5 million metric tons of seafood (Tacon et al., 2011). It has been predicted that by 2020, aquafeed demand may approach 71 million metric tons.

Historically, fish meal and oil were some of the most economical animal feed ingredients available. Over the past 10 years, however, the prices of fish meal and oil have skyrocketed. Reduction fisheries landings peaked at just over 30 million metric tons in 1994, but subsequently declined to reach a relatively steady state at approximately one-half that figure. Declining supplies and increasing demand over this period led to steep increases in the price of both fish meal and oil. From the mid-1980s to the mid-2000s, fish meal prices oscillated between approximately US$375 and US$750/metric ton; more recently, prices have fluctuated between seasonal lows of US$1300–1400/metric ton and highs of US$2000–2400/metric ton (World Bank, 2016a) (Figure 14.1). Similarly, fish oil prices hovered around US$400/metric ton for much of the 1980s and 1990s, and failed to top US$800/metric ton until 2006; recent

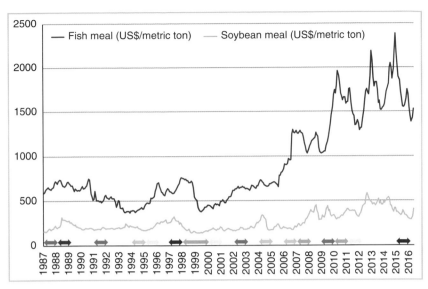

Figure 14.1: Prices of fish meal and soybean meal for the past 30 years. Arrows indicate the approximate occurrence and relative strength (color intensity is proportional to event strength) of El Niño (red) and La Niña events (blue).

Source: Data adapted from World Bank (2016a).

prices have ranged from US$1600–2400/metric ton (Food and Agriculture Organization, 2016a).

Even in times of relative stability, the prices of fish meal and oil are much more volatile than other commodities, such as corn, soy, and wheat. Reduction fishery landings vary from year to year like any wild catch, but the fish meal and oil markets are particularly vulnerable to the effects of ENSO events on oceanic productivity. The Peruvian anchovy fishery is the largest single source of fish meal and oil in the world. Anchovies are filter feeders, and in the late winter and early spring, the Peruvian fish crowd the coastline in huge schools to spawn and feast upon burgeoning phyto- and zooplankton populations. Thick plankton blooms occur along the Peruvian coast when strong currents bring cold, nutrient-rich waters from the dark ocean depths into the light and growing warmth of spring in the Southern hemisphere. When this upwelling occurs, plankton and anchovy are abundant. El Niño events sap the strength of the currents and upwelling, meaning that the cold water and nutrients stay far below the surface, beyond the reach of the strongest rays of sunlight, and the anchovy go hungry. To protect the weakened population from overfishing, quotas are greatly reduced in response to strong El Niño events. Of course, the market typically responds to the reduction in fish meal and oil supplies with temporary, but nonetheless influential price increases.

When commodity prices increase so dramatically, consumers who are unwilling to pay the high prices seek alternatives; the product volume they would have otherwise purchased becomes available and the increase in supply triggers a price reduction. This market correction is governed by the so-called "price elasticity of demand", or how much less (or more) of a product consumers are willing to purchase if there is an increase (or decrease) in price. Although they are not precisely interchangeable, ingredients such as corn gluten meal, soybean meal, and other plant protein sources are comparable in terms of their nutritional value, so if the price of corn gluten meal increases, buyers will simply switch to whichever of these ingredients is cheaper. These are elastic ingredients. For aquafeed makers, fish meal and oil are considered relatively inelastic: price hikes have relatively little effect on the demand for either ingredient. The same was true for poultry feed makers in the 1970s. At the time, most of the world's fish meal supply was being used in feeds for chickens, turkeys, and other fowl. Feeding soy and corn instead of fish promised significant cost-savings, but attempts to replace fish meal in poultry diets had largely failed to that point. Only after a concerted research effort to identify the unique attributes of fish meal and the limiting factors of other ingredients were nutritionists able to formulate fish-free feeds for poultry. Once the bounds of nutritional requirements and feed formulation were defined, fish meal became an elastic ingredient for the poultry industry. When the price per unit protein was economical, fish meal could still be used, but poultry feeds were no longer reliant on the ingredient. When the price climbed too high, feed manufacturers could simply reformulate their feeds and buy cheaper ingredients (Rumsey, 1993). Through research to define dietary demands and tolerances, some aquaculture sectors have been able to achieve the same flexibility: for example, tilapia, catfish, and carp will readily consume fish meal-based feeds, but can grow just as well – or at least as economically – on properly formulated fish meal-free feeds. For most carnivorous species, however, fish meal and oil remain inelastic and difficult to replace.

No fish *requires* fish meal or fish oil, but many require certain nutrients that are commonly found in marine-derived ingredients and few others. For example, fish meal is one of the most practical sources of taurine, a nutrient needed by obligate carnivores, including felines and a growing list of fish (Salze and Davis, 2015). Similarly, fish oil is the most economical source of long-chain polyunsaturated fatty acids arachidonic acid (ARA), eicosapentaenoic acid (EPA), docosahexaenoic acid (DHA), and others that play critical roles in the development, health, and well-being of humans and other vertebrates, including fish (Trushenski and Bowzer, 2013). For species that do not require these nutrients, replacing fish meal and oil begins with the relatively straightforward process of balancing the formulation to ensure the modified formulation does not contain too much or too little of other key constituents. Assuming the effects of alternative ingredients on the palatability of the feed (many carnivores exhibit aversion to the taste of plant-based feed) and the presence of "anti-nutritional factors" found in many plant-based ingredients[3] can be overcome, a fish-free diet can be implemented. This is how chickens, catfish, and others have been able to kick the fish meal habit. For species that do need taurine, ARA, or any of the other nutrients usually found in marine ingredients to live, alternative sources must be found. These nutrients can be supplied by synthetic sources (taurine) or specialty ingredients (EPA, DHA, and ARA can be found in algal oils), but these ingredients are expensive. Sometimes the cost of replacing fish meal or oil is higher than the cost of *using* fish meal or oil, so reverting to marine ingredients is more convenient and cost-effective.

---

[3] Anti-nutritional factors include several classes of compounds intended to protect plants from overgrazing. The effects of these compounds can be relatively mild (such as having an unpleasant flavor or interfering with nutrient digestion or absorption) or severe (causing allergic-like responses or toxicity). Some, but not all anti-nutritional factors can be attenuated by processing (for example, heat treatment, extraction and purification, fermentation).

Efforts to spare or replace fish meal and oil in aquafeeds are primarily driven by economic and logistical concerns. Rising feedstuff costs, especially marine ingredients, represent a significant threat to profitability for many aquaculture businesses. Given that aquaculture already consumes the vast majority of fish meal and oil supplies, future demand for aquafeeds cannot be met using existing formulations. Other justifications for limiting or eliminating fish meal and oil from aquafeeds are occasionally offered, some of which are legitimate. For example, the use of marine-derived ingredients has been linked to the accumulation of heavy metals and other contaminants in farmed fish. Though the human health risks associated with persistent organic pollutants and other toxic compounds in farmed fish are greatly exaggerated (see Chapter 6), reducing dietary exposure to these contaminants by adjusting dietary formulations can minimize the limited risk even further.

Other arguments for replacing fish meal and oil in aquafeeds are less cogent. Some have suggested that if aquaculture were to reduce its consumption of fish meal and oil, harvest pressure on reduction fisheries could be reduced, thus benefiting these stocks and strengthening the forage base for marine food webs. Any surplus is likely to be immediately consumed by any of the other industries that historically used fish meal and oil, particularly if the surplus resulted in falling prices. Hypothetically, the aquaculture industry could incentivize overfishing by allowing the price of fish meal and oil to climb because of price-demand inelasticity, but demand for reduction fishery products would exist with or without aquaculture. Regardless of whether aquaculture influences harvest pressure, reduction fisheries are aggressively managed and are considered sustainable. In fact, reduction fisheries may support modest growth in the future, particularly in La Niña years (Food and Agriculture Organization, 2014).

Others have argued against feeding fish to fish, whether as forage fish or in the form of fish meal or oil, claiming the practice is unnatural, particularly in the case of herbivorous fish. This argument is undercut by the fact that all fish begin life as ruthless hunters, stalking and eating zooplankton. Carnivory, including cannibalism, is part of the early life history of fish. Some fish adopt vegetarian feeding habits later in life, but most do not. Most continue to eat animals, including other fish. This is not to say that cannibalism has not led to catastrophic consequences in some terrestrial animals. Although "mad cow disease", more properly bovine spongiform encephalopathy, is likely the best known of the transmissible spongiform encephalopathies, prion diseases are known to occur in sheep and goats, elk, deer, some species of antelope, as well as mink, cats, and, of course, humans. The unique resistance of prion proteins to normal sanitation practices and the vulnerability of healthy cattle exposed to infected bovine-origin feedstuffs resulted in mad cow outbreaks throughout the world, most notably the epidemic beginning in the United Kingdom in 1986. Some have suggested that feeding fish to fish is the same as feeding bovine-origin ingredients to cattle, but this is not an analogous circumstance. Cattle and their ruminant ancestors have spent the past 50 million years, give or take, adapting to an exclusively herbivorous diet. The incorporation of animal-origin ingredients into the bovid diet – particularly meat, blood, and bone meals rendered from other cattle – was a decidedly unprecedented and artificial introduction. To find an instance of carnivory in fish, even among herbivorous species, one only has to go back as far as their larval stages. To discover a truly piscivorous common ancestor, one does not have to go back too far into evolutionary history: the herbivorous tilapias, for example, diverged from the other cichilds of the East African Lakes, including many carnivorous and piscivorous species, only 10–15 million years ago. This does not mean that fish are invulnerable to infectious prion diseases, only that such vulnerability would seem catastrophic in the evolution of species that regularly engage in carnivory and cannibalism. Interestingly, fish do not appear to be susceptible to mammalian prion disease (Dalla Valle et al., 2008) and do not pass the infective proteins in their tissues to other species (Loredana et al., 2006). Recall, too, that fish meal and oil not consumed by the aquaculture industry would likely be used to feed poultry and swine: if arguments regarding whether feeding fish to fish is natural are considered valid, surely feeding fish to fish would be preferable to feeding fish to pigs and chickens.

One of the more persistent myths surrounding aquaculture is that, by feeding fish to fish, the aquaculture industry consumes more seafood than it produces. Some have argued that it would be substantially more efficient to consume forage fish directly. On its face, this argument seems compelling: why not eat anchovies, herrings, and other forage fish instead of using them to fatten the fish that we ultimately consume? Of course, the answer to this question is a bit more complicated. Despite being highly economical and more nutritious than terrestrial meats and other seafoods, most consumers do not find forage fish appetizing. The fish are small, bony, oily, and if consumers have the buying power to choose other types of fish or meat, they typically do. Lower-value fish are an important source of nutrition for the poor, and redirecting more of the reduction fishery catch to direct consumption is a laudable initiative that would benefit the needy among us (Tacon and Metian, 2013). However, it seems unlikely that large segments of these industrial fisheries will be rebranded as food fisheries. As noted before, if the aquaculture industry did not use fish meal and oil, the poultry and swine industries would. So, the question is not really whether it is efficient to feed fish to fish, but rather whether feeding fish to fish is more or less efficient than feeding fish to chickens, pigs, or other livestock.

To answer this question, we must examine two related figures: feed conversion efficiency and dress-out percentage. Feed conversion ratios (FCRs) are a measure of how efficiently an animal grows, and are simple ratios between the amount of feed offered to the amount of weight gained. For swine, the FCR is usually about 3, meaning 3 kg of feed for every 1 kg of gain. Poultry are more efficient, gaining roughly 1 kg for every 2 kg of feed they eat. Fish, because they do not expend energy to maintain their body temperatures, are the most efficient, capable of achieving 1 : 1 conversions. A 3.6-kg (8-lb) Atlantic salmon typically yields a 2.7-kg (6-lb) carcass as a headed/gutted "bullet". From this bullet, one can expect to get 2–2.3 kg (4.5–5 lb) of retail cuts or saleable products, such as steaks or skin-on fillets. This equates to a 75–80% dress-out percentage (Chef's Resources, n.d.). Assuming a relatively modest 2 : 1 FCR, raising an 3.6-kg salmon would take 7.2 kg of feed. The calculations sort out to approximately 3–3.5 kg of feed in for every kg of retail salmon cuts out. For swine, the figures are a little different. A

113.6-kg (250-lb) hog yields a carcass weighing around 81.8 kg (180 lb). From this carcass, the dress-out to boneless chops and roasts is 52–65%, or 42.3–53.6 kg (93–118 lb) of retail cuts (Rentfrow, 2010). Assuming a FCR of 3, that comes out to more than 6–8 kg of feed in for every kg of retail pork cuts out. Chickens yield carcasses of about 75% of their live weight (Lessler and Ranells, 2007), which in turn yields approximately 70% edible meat and skin (Denton and Mellor, n.d.). A 2-kg (4.5 lb) bird would consume roughly 4.1 kg (9 lb) of feed and yield about 1.1 kg (2.4 lb) of edible products, meaning the feed to edible chicken ratio would be approximately 3.75. Resource utilization is measurably more efficient in production of farmed seafood than in terrestrial animal agriculture (see Chapter 16 for further details).

In terms of both FCR and dress-out, feeding fish to fish appears to be a more efficient means of producing food than feeding fish to swine or chickens. However, the misconception that aquaculture consumes more fish than it produces persists and is perpetuated by the misuse of "fish in, fish out" (FIFO) ratios. FIFO ratios measure efficiency by comparing the amount of fish meal and oil used in feeds in live weight equivalents to the amount of fish that is produced. On the surface, this seems like a good way to measure one of the environmental impacts of aquaculture. Unfortunately, the calculation is flawed, particularly when it comes to salmon. Salmon are fatty fish and consume diets that contain high levels of fats and oils, including fish oil. Based on the amount of fish meal and oil they are fed, the standard FIFO ratio for salmonids is 2.5 – 2.5 kg of wild fish in for 1 kg of farmed fish out. The problem is that these calculations do not account for the fact that the rendering process yields much more fish meal than fish oil. So, it may take 2.5 kg of wild fish's worth of fish oil to raise 1 kg of farmed salmon, but it does not take nearly that much fish meal. The leftover fish meal is fed to other fish (or terrestrial livestock), yielding additional farmed fish that is not reflected in the salmon FIFO ratios. Using a more mathematically correct calculation that takes this into account, the corrected FIFO ratio for salmonids is actually 1.27 (Jackson, 2013). Naturally, the corrected figures for species that consume less fish meal and oil are even lower. In 2014, approximately 21 million metric tons of fish were harvested for industrial purposes (76% of this volume was rendered to fish meal and oil) and approximately 51 million metric tons of fed aquaculture species were produced (Food and Agriculture Organization, 2016b). Even assuming the entire industrial fisheries catch were devoted to feeding farmed fish (again, less than 76% was in 2014), the global FIFO ratio would be, at most, 0.4 (for 1 kg of fish in, 2.5 kg of fish out). One can argue that raising mostly herbivorous fish and directly consuming more forage fish would increase efficiency, but these points are largely moot, as this is what the global aquaculture industry does already.

How the aquaculture industry uses fish meal and oil is a subject worth examining: it is critical that these highly valuable, but limiting resources be used judiciously to ensure aquaculture's long-term economic sustainability. Inclusion rates for fish meal and oil have been steadily declining for years and their use is being increasingly shifted to specialty diets for larvae, brood fish, and other key life stages (Tacon et al., 2011). These efforts, bolstered by intensive research to identify limiting nutrient requirements and develop alternative ingredients for aquafeeds, will continue. That said, it is likely that fish meal and oil will always be used by the aquaculture industry. In light of the truths of feeding fish to fish, there seems little wrong with that.

PART IV

Socioeconomic and related issues

Chapter 15

Economic interactions between farmed and wild-caught seafood

A rising tide lifts all ships. (Various attributions, unknown)

Throughout most of the world, terrestrial animal protein is produced by farmers and ranchers who face little competition in the marketplace from hunters. Beyond the environmental effects of animal agriculture on habitat and environmental quality, there is little to discuss regarding the interactions between livestock and wildlife. Headlines do not commonly ask if the benefits of cattle ranching outweigh its negative impacts[1] or whether poultry are a farmed and dangerous threat to game birds or waterfowl.[2] Elsewhere in this title we have explored the unique onus of being the "last wild food" as well as many of the ecological interactions between farmed and wild fish. But what of the invisible hand of the market? How do wild and farmed fish interact in the marketplace and how are these unseen market forces responding to the growing availability of farm-raised seafood?

The financial crisis of 2007–2008 and the subsequent "Great Recession" brought some of the bewildering complexities of economics and the modern financial system into the public consciousness: imbalances and bubbles, credit default swaps, shadow

---

[1] "Do the Benefits of Aquaculture Outweigh Its Negative Impacts?" (SENCER, 2017).
[2] "Farmed and dangerous? Pacific salmon confront rogue Atlantic cousins" (Braun, 2017).

banking. Most economists failed to predict the coming crisis, in part because even they did not fully understand the financial practices and products that ultimately led to the global economic downturn in the late 2000s. The economics of global seafood trade are also complex, but the economic interactions between farmed and wild fish in the marketplace are relatively straightforward. They mostly have to do with the core economic principles of supply and demand that even those of us with the least economic prowess can understand. In the simplest of terms, the price that consumers are willing to pay for a product or service is greatest when availability is limited and least when there is an abundant supply. In short, one can expect prices to fall as goods or services become more available (Figure 15.1).

Capture fisheries landings have not changed appreciably for decades, but the burgeoning aquaculture industry has made seafood increasingly available. Aquaculture's ascendancy is itself a response to market forces: poor fisheries management, inefficiency in the fishing sector, and the collapse of many once-lucrative fisheries created the supply vacuum in which aquaculture has flourished (Anderson, 2007). Applying the classical understanding of supply–demand relationships, one would assume that seafood prices declined sharply as the supply of farmed fish grew. As it happens, this is precisely what occurred to the Alaskan salmon market during the last 20 years of the 20th century: as farmed fish became common in the marketplace, the value of salmon fell to one-third of 1980s prices (Knapp, 2007). Previously insulated from competition, the Alaskan fishing industry was able to persist profitably despite product quality issues and a fleet bloated with many more boats and fishermen than needed to bring in the catch. These problems could not be ignored once farmed salmon presented consumers with a choice. Summarizing the Alaskan experience, economist Gunnar Knapp acknowledged there were two perspectives on the growth of salmon aquaculture and the consequences for salmon fishing. The first, most commonly echoed by the fishing industry, alleged that farmed salmon, "… flooded world markets, depressing wild salmon prices and significantly harming Alaska fishermen and fishing communities" (Knapp, 2007: 245). Knapp's own, more conciliatory perspective

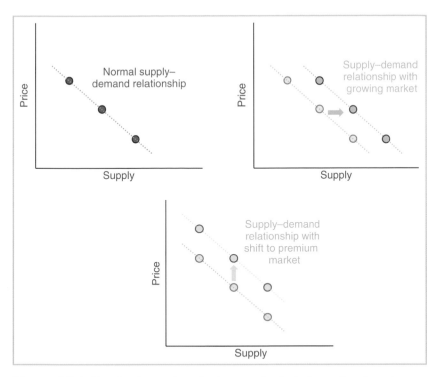

Figure 15.1: Supply, demand, and price are classically interrelated: under normal circumstances, as a product or service becomes increasingly available (that is, supply increases), consumers' desire for the product/service (that is, the price they are willing to pay) declines. In the case of Alaskan salmon, increasing supply from aquaculture caused prices to drop in the 1980s and 1990s, negatively affecting the fishing sector. In other cases, aquaculture has shifted the supply–demand curve to the right by creating new markets or upwards by allowing niche marketing of premium products. Where aquaculture has helped to grow the consumer base or pushed the fishing industry to improve and differentiate their products, increasing supply of farmed seafood has not had the predictable effect on product pricing or the profitability of capture fisheries.

was that, "Salmon farming exposed a 'natural' monopoly to competition, benefiting consumers by expanding availability, lowering prices, spurring innovation and market development, and leading to a more efficient wild salmon industry more focused on meeting market demands" (Knapp, 2007: 245).

Increasing supply from aquaculture has had a similar effect on seafood prices in other sectors, but it has also strengthened product pricing and created new economic opportunities. Certainly, strong demand for fish meal and oil from the aquaculture industry has helped the value of reduction fisheries to increase over the past 15–20 years (Tacon and Metian, 2008), but has not caused fishing pressure or harvests of forage fish to increase (Asche and Tveteras, 2004).[3] Aquaculture has also created opportunities for wild and farmed food fish. By making lower-cost seafood products available, the aquaculture industry has encouraged consumers to give seafood a try, growing the consumer base and expanding markets (Tveteras et al., 2012). Although seafood demand is undeniably influenced, first and foremost, by population growth, increasing per capita consumption is also growing markets for both wild and farmed fish (Figure 15.2).

Aquaculture is also growing seafood markets by introducing consumers to new products. For example, catfish and tilapia were virtually unknown to American consumers until the aquaculture industry made these products available (Anderson, 2007). In this way, aquaculture has created entirely new markets for both farmed and wild fish. The availability of lower-cost, farm-raised seafood has also allowed established markets to differentiate: in the past, a salmon was a salmon was a salmon; but today there are commodity markets for consumers primarily interested in affordability (pink salmon and chum salmon) as well as niche markets for those who are willing to pay a premium for perceived quality and "wildness" (Chinook salmon, coho salmon, and sockeye salmon) (Anderson, 2007). In fact, the strength of these marketing efforts has allowed harvest to continue virtually unchanged in many fisheries, despite growing availability of an equivalent, lower-cost farmed product (Naylor et al., 2000). Returning to the Alaskan example, fishing effort for salmon did not drop off in response to declining prices in

---

[3] From a strict supply–demand perspective, increasing demand for fish meal and fish oil would be expected to increase harvest pressure, but reduction fisheries are governed by more than market forces and supply–demand dynamics. Today, regulations limit harvest and protect these fisheries from overharvest (Asche and Tveteras, 2004).

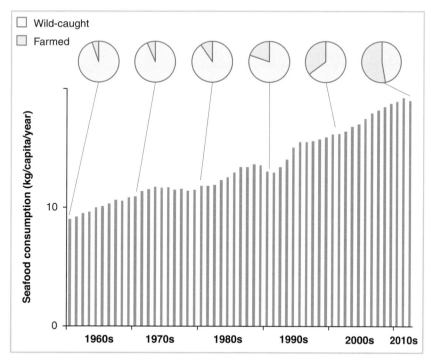

Figure 15.2: Although global seafood demand increases primarily as a function of human population growth, increases in per capita consumption also have an effect. Per capita consumption began to increase rapidly in the late 1980s, as aquaculture made seafood increasingly available and affordable. Data adapted from FAOSTAT (Food and Agriculture Organization, 2017l).

the late 1980s and 1990s – global salmon catches increased by more than 25% during this time.[4]

Aquaculture products have generally not replaced wild-caught fish and shellfish in the marketplace, but grown the seafood market as a whole; indeed, a rising tide does lift all ships. The aquaculture industry has undoubtedly changed supply and demand for seafood and the opportunities for wild-caught products, but not necessarily

---

[4] That aquaculture has generally not displaced fisheries cuts both ways: critics use this as evidence that aquaculture is not able to relieve harvest pressure on wild fish populations, but some would also lament the loss of traditional livelihoods if capture fisheries were to contract as a result of competition with aquaculture.

for the worse. More and more of the seafood we consume will come from aquaculture, and factors that affect production costs and product value in aquaculture will become more influential as the global seafood trade becomes increasingly tied to fish farming (Kobayashi et al., 2015).

In aquaculture, the primary costs are fingerlings and feed – there is tremendous opportunity to reduce both costs through research and development in fish nutrition and genetics. In capture fisheries, the primary costs are labor, fuel, and maintaining the fleet – innovation may reduce costs here, too, but there is substantially less room for cost-savings in these sectors and the fishing industry is not especially influential in any of them (Anderson, 2007). Based on these variables, it would seem that aquaculture is destined to become more economical than fishing, but disease outbreaks, shifts in consumption patterns, or changes in the pricing of fish meal or other commodities used in feed manufacturing could quickly derail such progress (Kobayashi et al., 2015). The cost of a fillet may become increasingly dependent on whether the fish was caught or raised, but whether farmed or wild fish will be more affordable in the future depends on the movements of many invisible hands.

Chapter 16

Resource utilization in the production of animal protein – seafood versus other meats

We must plant the sea and herd its animals using the seas as farmers instead of hunters. (Jacques Cousteau, French explorer, conservationist, and world-famous marine biologist)

The world is hungry and increasingly hungry for protein. There are just over 7 billion of us on the planet at this time, and over the next three decades or so, it is estimated that our ranks will swell to more than 9 billion. By next century, the human population may be nearly 11 billion strong. Based on population growth, we will need to be producing 60% more food by 2050. Not only will we need more food in general, we will need a lot more animal protein in particular. Increasing urbanization, lower production costs, greater buying power, and the like have led to increases in total per capita consumption of animal protein. The protein supply has steadily increased from about 61.5 g/capita/day in the 1960s to more than 81 g/capita/day in 2013, with the increase coming largely from an increased supply of protein from animal products (Figures 16.1 and 16.2). These increases – while impressive – are not enough. We need 60% more *food* by 2050, but we need 60% more *animal protein* by 2030 (CGIAR, n.d.).

Producing this much more animal protein is going to require a significant amount of inputs, regardless of whether it comes from terrestrial animals or seafood. As was discussed in Chapter 14, aquatic animals convert feed resources into food resources more efficiently than terrestrial livestock: dress-out percentages tend to favor aquatic

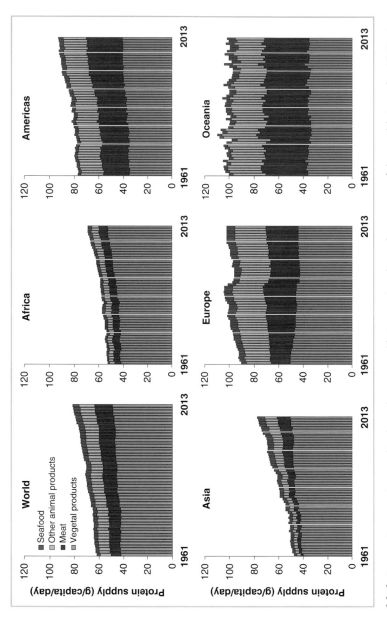

Figure 16.1: Per capita protein supplies have been steadily increasing throughout most of the world since the 1960s. In most cases, the increase has resulted from increasing production and consumption of animal products (Food and Agriculture Organization, 2017m).

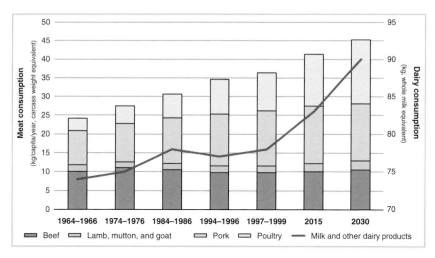

Figure 16.2: Consumption of meat and other animal products has been steadily increasing throughout much of the world, but the trend does not apply to all types of animal products equally. Consumption of beef, lamb, mutton, and goat meat has changed little since the mid-1960s, whereas pork and poultry consumption have increased substantially. Dairy consumption has also increased, particularly since the mid-1980s (Food and Agriculture Organization, 2003).

animals because of their body conformations and, because fish and shellfish do not expend energy to regulate their body temperatures, more of the energy they consume is directed to growth, meaning greater feed conversion efficiencies. As discussed in Chapter 10, water usage in aquaculture is often non-consumptive and less per unit food produced than other forms of livestock production. Of course, there is more than just water and feed or forage that goes into raising an animal. How does one begin to examine all of the inputs and outputs of raising terrestrial and aquatic animals and decide which is the most efficient means of doubling world meat production by 2050 (Food and Agriculture Organization, 2016j)?

Life cycle analysis (LCA) – a means of quantifying resource use associated with products or services – is one way of answering this question. The methods vary considerably, but all LCAs intend to describe the environmental effects of an activity, capturing the use of energy and materials (inputs) as well as wastes and environmental consequences (outputs) from "cradle to grave" or, in the case of food, from "field to fork." For assessments of animal agriculture, inputs might include such things as feed or fertilizer resources, fuel for vehicles or implements used in production, the physical space needed to raise the animals, and so forth; outputs might include emissions of greenhouse gases from the animals or combustion of fossil fuels, eutrophication caused by nutrient-rich effluents, and the like. The results of LCAs – carbon footprints, land or water needed to produce a unit of the relevant product, or similar metrics – can be used to compare products or production methods in terms of their sustainability, effects on the environment, and so forth.

There are a number of factors that can influence the land use or carbon footprint of animal agriculture operations. Of course, large terrestrial animals take up more space than small or aquatic animals, but land use has more to do with the amount of arable land used to produce the animals' feed. For example, it takes a good deal more land to produce 1 kg of free-range beef than to produce the same amount on a cattle feedlot in part because the feedlot takes up less space than the rangeland, but mostly because grassland cannot produce nearly as much cattle fodder as a cornfield can. Even though the land requirements for industrially produced beef must include the size of the feedlot and the farmland used to produce the corn and other fodder, LCA typically indicates that feedlots are more efficient than rangeland. In fact, in terms of land use, free-range cattle offer the least efficient means of producing meat or protein. Farmed and wild-caught seafood are at the other end of the spectrum, requiring little-to-no arable land to produce protein for human consumption. Land use in aquaculture – including the physical space occupied by inland farms as well as arable land used to produce feed ingredients – is generally comparable, but usually less than that used in poultry, pork, and other terrestrial livestock sectors (Figure 16.3).

Carbon footprints are a bit more helpful than land use in drawing comparisons between terrestrial and aquatic animal protein production. Carbon footprints are usually quite high for cattle, sheep, goats, and other ruminants because of methane they produce during rumination and the nitrous oxide generated by manure management.

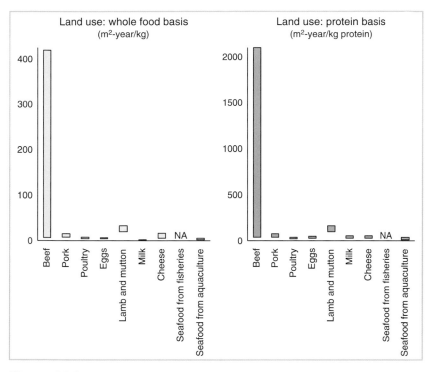

Figure 16.3: Land use in animal protein production varies considerably, from zero in the case of capture fisheries, to as much as 2100 m²-year/kg protein in the case of free-range cattle reared on open grasslands. For those less mathematically inclined, that is about 150 McDonald's quarter pounders per year from a football field's worth of grassland. Of course, there is little land used directly in aquaculture, but the figures shown here include the land used to produce terrestrial crops used to make aquafeeds (Nijdam et al., 2012).

Of the roughly 7 gigatons[1] of CO_2 equivalents of greenhouse gases generated annually by livestock production worldwide, more than 80% is attributable to ruminants (Gerber et al., 2013). Beef (41%) and dairy cattle (20%) are the largest contributors to greenhouse gas emissions, followed by buffalo (9%) and small ruminants (6%) raised for meat and milk, pork (10%), and chicken raised for meat (6%) and eggs (3%) (Figure 16.4).

---

[1] A gigaton is one billion metric tons. For reference, 7.1 gigatons is equivalent to the weight of about 35 million blue whales – the largest living animals on the planet – or more than 1000 Great Pyramids of Giza.

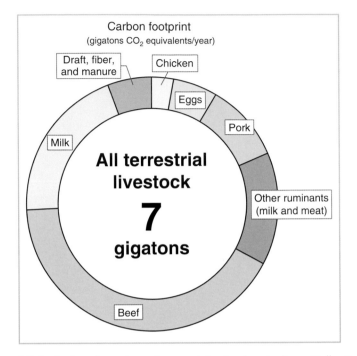

Figure 16.4: Carbon footprints of terrestrial animal agriculture. Collectively, livestock are responsible for about 14.5% of human-related greenhouse gas emissions. Equivalent statistics are not readily available for fisheries or aquaculture, but the smaller size of these sectors and their smaller carbon footprints per unit protein suggest that the greenhouse gas emissions associated with wild-caught and farmed seafood are probably insignificant in comparison (Gerber et al., 2013).

Most of the ruminant carbon footprint is associated with raising and harvesting crops as feedstocks and the biological processes of cattle themselves, though handling of manure and the resulting milk and meat also contribute. Feed is the most important element of the poultry and swine footprints, though manure management is an important source of greenhouse gases for these species also. It is difficult to identify comparable estimates for the fishing and aquaculture sectors, but their smaller scale and operational needs contribute to much smaller carbon footprints. For example, it is estimated that fuel usage by the global fishing fleet generates 0.000134 gigatons of CO_2 annually (Muir, 2015), but this figure does not capture energy costs and emissions associated with processing, packaging,

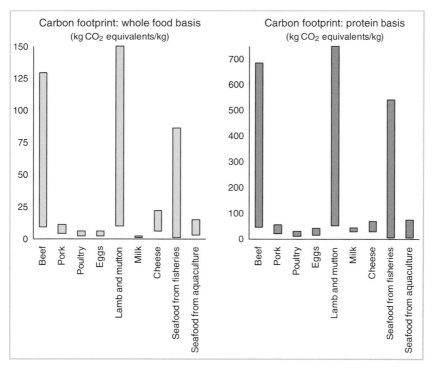

Figure 16.5: One of the most commonly used metrics used in LCA is carbon footprint, expressed in terms of greenhouse gas emissions, such as methane produced by ruminants during digestion of their forage or carbon dioxide resulting from animal respiration or the combustion of fossil fuels during operation of vehicles, implements, or vessels. Seafood produced in aquaculture has a relatively small footprint, but fuel consumption drives up the footprint of wild-caught seafood (Nijdam et al., 2012).

and transportation of seafood products, which can be significant.[2] Although global estimates are lacking, carbon footprints per unit whole food or per unit protein can shed some light on the efficiency and environmental consequences of satisfying protein demand with seafood versus terrestrial meats (Figure 16.5). Of course, ruminant animals have the largest footprint per kilogram of meat or kilogram

---

[2] Carbon footprints are much larger for seafood that is transported by air. Although locally caught and consumed seafood generates little in the way of CO_2 emission equivalents, distribution by airfreight may generate as much as 8.5 kg of CO_2/kg fish shipped (Food and Agriculture Organization, 2008).

of protein – increasing the protein supply by raising more cattle, buffalo, sheep, and goats could have catastrophic consequences for the environment and efforts to mitigate human contributions to climate change. As was the case for land use, other sources of animal protein, like poultry, pork, and farmed seafood, have substantially lower carbon footprints – the exception is seafood from capture fisheries, which may have carbon footprints as large as those reported for ruminants. Fuel usage varies considerably among fisheries (Parker and Tyedmers, 2014), and depending on how remote the fishing grounds are and how far the fish have to travel before they reach consumers, the carbon footprints can become quite sizable.

In the next 30 years or so, humans must address a serious shortfall in animal protein production. We will need to roughly double animal protein production in order to satisfy the demands of more

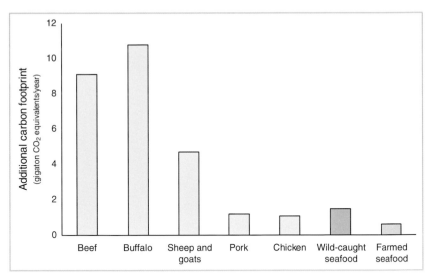

Figure 16.6: Approximately 200 million metric tons of additional animal protein will be needed each year by the year 2050. The shortfall will undoubtedly be met by growth in a number of livestock sectors, but what would the carbon footprints of 200 million metric tons of beef, chicken, pork, or seafood look like? Based on emission intensities reported for terrestrial livestock production (Gerber et al., 2013) and selected capture fisheries and aquaculture sectors (Parker and Tyedmers, 2014), the carbon footprints would range from roughly 0.5 to 11 gigatons of CO_2 equivalents.

than 9 billion people living on Earth. Each year, roughly 200 million metric tons of additional meat and other animal products will be needed. The poultry and pork industries are expected to grow worldwide, but ruminant production is expected to hold steady for the next few decades. No single sector will be responsible for closing the gap, but what if they did? What would those carbon footprints look like? An additional 200 million metric tons of pork or chicken would generate an additional gigaton of CO_2 equivalents – beef or buffalo would come with about 9–11 gigatons (Figure 16.6). And what of seafood? The increase in greenhouse emissions would vary based on whether the fish was wild-caught or farmed, but it could be less than one gigaton of CO_2 equivalents per year.

Population growth may fall short of predictions. Dietary habits may trend away from increasing consumption of meat, eggs, milk, and seafood. The coming protein crisis may not be a severe as some would predict. But we will need more protein in the future, and it will come at an environmental price. Whether we choose to pay the lowest environmental costs and encourage sustainable aquaculture development and farming of the seas remains to be seen.

Chapter 17

Social and economic empowerment through aquaculture

Women who are offered and provided with the best circumstances to enhance their socioeconomic empowerment will also be able to contribute meaningfully to food security, poverty alleviation and improved well-being for themselves, their families and their communities. In short, they will help to create a world in which responsible and sustainable use of fisheries and aquaculture resources can make an appreciable contribution to human well-being, food security and poverty alleviation. (Food and Agriculture Organization, 2012)

For those who enjoy the liberties and opportunities that come with living in the relative comfort of the industrialized world, the plight of disenfranchised women, the rural poor, and others with little personal agency or prospects may seem remote. For unacceptably large segments of the world's population, however, the struggle to achieve basic food security and economic empowerment is part of daily reality. Despite progress made to address poverty throughout the world, roughly one in ten people live below the extreme poverty line, subsisting on less than US$1.90 per day. Most of these 767 million people live in sub-Saharan Africa, South Asia, East Asia, and the Pacific, but the hunger, inadequate housing, and poor health they experience over the course of their comparatively short lives is not unique to these regions. Regardless of location, those living in poverty often have much in common: they are mostly young, uneducated, employed in the agricultural sector, and have large families with many children (World Bank, 2016b).

Poverty is directly linked to undernourishment. Economic development has done much to address hunger throughout the world. In last 25 years, the number of undernourished people has declined from more than 1 billion (or 18.6% of the population at the time) to 795 million (or 10.9%)[1] (Food and Agriculture Organization for the United Nations/International Fund for Agricultural Development/World Food Programme, 2015). Reducing the number of undernourished people by 216 million since 1990 is impressive progress, particularly when one considers that the human population grew by more than 2 billion during this time period (World Bank, 2017a). Unsurprisingly, undernourishment is most prevalent in developing regions (currently 12.9% of the population in these countries), but still exists in the developed world (<5%). Gains in food security and adequate nourishment have been regionally uneven: whereas undernourishment has decreased impressively in Asia (23.6 to 12.1%) and in Latin America and the Caribbean (14.7 to 5.5%), progress in Africa (27.6 to 20.0%) and Oceania (15.7 to 14.2%) is comparatively meager (Food and Agriculture Organization for the United Nations/International Fund for Agricultural Development/World Food Programme, 2015). The consequences of malnutrition extend throughout the lives of the chronically undernourished, affecting fetal and early childhood survival and developmental outcomes and limiting the physical well-being and mental capacity of adolescents and adults. The problem is compounded over generations, as the learning deficits of undernourished children lead to poor educational performance, fewer marketable skills, under- or unemployment, and reduced capacity to guarantee food security for their own families. Extrapolated across populations where undernourishment is widespread, the societal consequences become increasingly troubling (Tacon and Metian, 2013).

---

[1] Note how closely this figure aligns with the number of individuals estimated to be living below the extreme poverty line. Given the nature of these datasets, one might consider the difference between the 767 million living in extreme poverty and the 795 million people who are undernourished to amount to little more than rounding error.

Although various circumstances complicate gender-specific inter-pretation of global poverty statistics,[2] there are a number of factors that suggest women are disproportionately vulnerable to poverty and food insecurity. In male-led households, the nutritional needs of women may be considered subordinate to the needs of children or men, particularly male breadwinners. In both male and female-led households, opportunities for women to provide greater food security for their families may be constrained by a variety of impedi-ments or barriers. In many cultures, education of girls and women is underprioritized or actively discouraged. Consequently, of the roughly 780 million illiterate adults in the world, two-thirds are women. Opportunities for education and training are also limited for girls and women by their traditional roles in child-rearing and other household work. These responsibilities often require women and girls to be physically in the home and leave little time for other pursuits, making education and work outside the home difficult, if not impossible. In some regions, the ability to seek opportunities outside the home is further constrained by mobility. Until 2018, Saudi Arabia famously banned women from driving cars, and in many Islamic cultures, women are discouraged or prevented from appearing in public without their husbands or other acceptable male chaperones. Activities outside the home may be temporally limited, further restricting women's access to media and informa-tion and limiting their participation in public meetings, training programs, and the like. Women may also be culturally or legally prohibited from owning or inheriting property, meaning they have little opportunity to engage in the farming or grazing opportunities

---

[2] Poverty data are typically tracked by household, without any tracking of gender composition beyond the head of household. Although one can compare the living conditions and resources of households led by women versus men, the World Bank notes that such comparisons provide little value, since female heads of household reflect diverse circumstances. Women may be heads of households because they have the economic means to do so; because male family members are not physically occupying the home, but are providing support by securing employment in other locations; or because they are widows. It is likely that the economic standing of the households in these scenarios is quite different, making direct comparisons between male- and female-led households difficult.

that often form the backbone of rural economies. Although many of the cultural and legal prohibitions that prevent the involvement of women in public life are associated with Islamic fundamentalism and Sharia law, predominantly Muslim countries are not the only ones to legally restrict the free agency of women. For example, laws in China, Russia, and Madagascar prohibit women from engaging in certain types of work or from working at specified times. In Cameroon, husbands may legally object to their wives' work and prevent their participation in trades; in other countries, men are given de facto authority to prevent participation of women in the workforce by granting husbands the right to prevent their wives from leaving the home except with their permission (Equality Now, 2015; Strochlic, 2015). Legal roadblocks such as these are reinforced perceptions informed by social mores that are collectively more difficult to reverse than any particular law (Figure 17.1).

With less education, fewer career opportunities, and limited job and leadership experience as a result, women are globally underrepresented in the work force[3] and decision-making processes (United Nations Statistics Division, 2015). Given these various constraints on the ability of women to participate in the workforce and marketplace, they are disproportionately vulnerable to disenfranchisement and food insecurity:

> Pre-existing gender-based disparities in access to assets such as land, property or credit mean that women have often fewer financial resources than men to cope … while limited mobility and work opportunities outside the home reduce their range of coping strategies … with male household members absent, women are not always able to claim family assets, such as land, livestock, tools and machinery, previously owned by their husbands, especially if they are illiterate or insufficiently aware of their legal rights with significant negative implications for food security. (Food and Agriculture Organization/International Fund for Agricultural Development/ World Food Programme, 2015: 39)

---

[3] Fifty percent of working-age women are employed, compared to 77% of working-age men (United Nations Statistics Division, 2015).

Conservative	Less conservative
Women should not work outside the home because of social and religious reasons and in keeping with tradition	It is acceptable for women to work outside home
Having women working in the field will result in a poor harvest (superstition)	Women learn from working outside the home
Women have no time to work outside the home	Men appreciate and value women's work and skills
Women are unable to do many kinds of work	Both men and women are needed to manage a household

Conservative	Less conservative
Women generally do not want to work outside the home	Women want to work outside the home
Women may want to work outside the home, but there are no opportunities	Husbands are supportive and appreciate women's work
Husbands will not permit their wives to work outside the home	Women can work near the home with men
Women are unable to work outside the home or do not have time to do so	Women want to improve their family's welfare

Figure 17.1: Perceptions of men (in blue) and women (in pink) living in conservative or less conservative cultures regarding participation of women in agriculture, including aquaculture. Adapted from Gender in Agriculture Sourcebook (World Bank/Food and Agriculture Organization/International Fund for Agricultural Development, 2009)."

In this context of extreme poverty and food insecurity for roughly 10% of the world's population, growing demand for animal protein, and the challenges of integrating women and the impoverished among us into positions of security and agency, aquaculture can be an instrument for social and economic change.

Undernourishment encompasses inadequate caloric intake, as well as dietary deficiencies in macronutrients (such as protein or fat) or micronutrients (vitamins and minerals). In these circumstances, seafood is critical, particularly for many of the world's most vulnerable impoverished populations. Seafood is typically a better source than terrestrial meats for protein, limiting amino acids methionine and lysine, long-chain polyunsaturated fatty acids, most essential minerals and trace elements, and a variety of vitamins (Tacon and Metian, 2013), all of which tend to be underrepresented in the plant-based diets of the rural poor. Consequently, even in small amounts, seafood can significantly enhance the nutritional value and food security of those living in poverty and struggling to achieve or maintain food security. What is more, in many countries the average market price for fish is lower than that of poultry, pork, beef, and other terrestrial meats, and aquaculture production can further reduce the price of seafood (World Bank, 2007). Thus, seafood in general – and farmed fish and shellfish in particular – can help the limited monetary resources of the poor to stretch a bit further in satisfying their dietary needs. In this light, it is perhaps unsurprising that 18 developing countries in sub-Saharan Africa and 14 in Asia derive the majority of their animal protein supply from fish and shellfish (Tacon and Metian, 2013).

Aquaculture can be a subsistence activity, providing needed food for the impoverished and undernourished, but it can also be a means of escaping poverty. Whereas significant economic barriers block entry to other forms of livestock production, extensive aquaculture requires little more than seedstock and access to water to begin. Enabling policies can further ease entry into aquaculture, such as measures that allow low-cost leasing of public waters or unusable land (World Bank, 2007). Those employed in aquaculture may also receive greater compensation than those involved in other forms of agriculture. For example, Chinese farmers raising catfish and shrimp may earn $1000 annually, roughly four times the household income provided by terrestrial farming. Similarly, tilapia farmers in the Philippines may earn nearly 50% more than rice farmers, and Bangladeshi farmers producing shrimp and rice in integrated paddies may double their farm's gross returns (World Bank, 2007). Collectively, examples from Asia suggest that even among those who do not escape poverty, aquaculture substantially increases quality

of life and food security. There is some suggestion that incorporation and centralization of aquaculture in the developing world may
increase production efficiency, but limit benefits to the rural poor.
By creating aquaculture cooperatives, small-scale producers can take
advantage of collective buying power and economies of scale, while
keeping the majority of decision-making and profits at the individual farmer level.

Collectively, fisheries and aquaculture are estimated to provide jobs or revenue to roughly 200 million people (World
Bank/Food and Agriculture Organization/International Fund
for Agricultural Development, 2009). Approximately 18 million
people are engaged in aquaculture worldwide, and this figure has
held relatively constant since 2005. Women represent 18% of
those engaged in the primary business of raising fish and shellfish, but are more greatly represented in processing and other
support sectors; when these industries are factored in, women represent half of the aquaculture workforce (Food and Agriculture
Organization, 2016b).

These statistics are encouraging, but there are nonetheless vital
differences in the power positions of men and women in fisheries
and aquaculture. In general, women have less control over the value
chain, their activities are not compensated as well, and they may
be actively or passively excluded from more profitable markets and
higher-paid positions. For example, in the shrimp farming industry,
women are largely restricted to the lowest paid and most vulnerable
segments of the workforce (for example, fry catchers, laborers, and
low-paid processing positions) whereas men tend to occupy the more
lucrative and stable positions (for example, farmers, middle men,
and exporters). What is more, women often receive lower wages
than male counterparts engaged in the same work: in Bangladesh,
female fry catchers may only receive 64% of the wages given to male
fry catchers. Wage gaps and inferior working conditions for women
are not unique to Bangladeshi aquaculture, as similar circumstances
have been reported for segments of the aquaculture industry in India,
Kenya, Sri Lanka, Latin America, and elsewhere (World Bank/Food
and Agriculture Organization/International Fund for Agricultural
Development, 2009).

Gender differences in compensation notwithstanding, for rural women, aquaculture presents a relatively unique opportunity to produce animal protein and generate income. As noted above, the initial investment needed to begin subsistence aquaculture is relatively low. What is more, many forms of aquaculture can be practiced closer to home and with less active time investment than, say, livestock grazing. Aquaculture is recognized as a means of increasing livelihood security for the rural poor and addressing the burdens of food scarcity and poverty that disproportionately affect women in these regions. Nongovernmental organizations have actively promoted "family based" approaches to aquaculture in places where social or cultural barriers limit the ability of women to engage meaningfully in work outside the home. Bangladesh has been a specific target for development of family based aquaculture, whereby both men and women are trained and participate in raising fish or shellfish for familial subsistence or for sale. Gains have been made, but clearly there is more work to be done to facilitate gender mainstreaming in aquaculture, particularly in places that are considered to be socially conservative (World Bank/Food and Agriculture Organization/ International Fund for Agricultural Development, 2009).

It is not hyperbole to say that aquaculture has the opportunity to improve and save lives. Vitamin A deficiency is a serious problem in Bangladesh, where an estimated 60% of the population fail to consume enough of this essential nutrient to prevent blindness, childhood mortality, and other medical complications of vitamin A deficiency. Increasing small-scale production of a small vitamin A-rich fish, the mola carplet, to a relatively modest 27,374 metric tons per year[4] has been suggested as cost-effective means of increasing vitamin A intake among Bangladeshis. If implemented, increased mola carplet culture could save as many as 3000 lives over an 11-year implementation period, many of them women and children (Fiedler et al., 2016). Whether in this narrow context of addressing vitamin

---

[4] Bangladesh is one of the largest aquaculture producers in the world, raising nearly 2 million metric tons of mostly freshwater fish annually (Food and Agriculture Organization, 2016b). The proposed increase in mola carplet aquaculture would increase annual Bangladeshi aquaculture production by about 1.5%.

A deficiency in Bangladesh or the broader context of increasing food security, economic opportunity, and personal agency for the most vulnerable among us, aquaculture can be transformative. Perhaps it is time to reimagine the proverb about teaching a man to fish to more accurately reflect the modern era: give a man a fish and you feed him for a day; teach men and women to raise fish, and you feed them, their families, and the world for lifetimes.

Chapter 18

Carp or salmon? Meeting seafood demand in developed and developing nations

Smoked carp tastes just as good as smoked salmon when you ain't got no smoked salmon. (Patrick McManus, American author and humorist)

In Chapter 3, we discussed the staggering diversity of aquaculture. It is clams and cod, kelp and koi, shrimp and sturgeon. Aquaculture is more than carp and salmon, and yet these two types of fish frame many discussions about the ways, means, and ethics of raising fish and shellfish. Carp are the fish of the common man – a proletarian source of protein, feeding low on the food chain. They transform what little food they are given into thickly muscled bodies, thriving in conditions that would choke and poison other less-robust species. Salmon are more fastidious, more discriminating – a food for the aristocracy. To reach their full potential, salmon must have pristine water and a carnivorous diet nutritionally superior to that of many of the world's poor. When juxtaposed, carp and salmon illustrate two very different types of food, two different types of aquaculture. Omnivore versus carnivore. Dietary staple versus luxury food. Low versus high market value. Both carp and salmon are delicious and provide essential nutrients, but the cost of salmon puts it out of many consumers' reach.

Aquaculture is growing fastest in the developing world, where physical and economic access to adequate nutrition – food security – is a daily concern. Aquaculture can do tremendous good in such places, providing a stable source of needed protein as well as economic opportunity (see Chapter 17). One might assume that production in the developing world would tend to favor carp and other easily propagated species. To a certain extent this is true: extensive aquaculture of herbivorous and omnivorous species is prevalent throughout much of the developing world. Indian major carp, Asian carp, and other low-trophic level species are the most important cultured fish in India, China (still considered a developing country by some economic measures), Indonesia, Bangladesh, and other developing countries (Food and Agriculture Organization, 2018a, 2018b, 2018c, 2018d). These countries also consume most of their freshwater fish production domestically, helping to address the food security needs of their citizenries.

Many of these countries are also raising high-value species. Shrimp raised in Ecuador and Vietnam are not destined for Ecuadoran and Vietnamese tables. Groupers raised in Malaysia are not intended for Malaysians. Cobia raised in Panama are not prepared Panamanian-style. These high-value species are exported to the United States, Japan, and other regions where consumers can afford to pay top-dollar. Although this strategy contributes to the gross domestic product of the developing countries, it does not directly feed their people. Although the developing countries are compensated for the seafood they export, there is something discomforting about impoverished farmers selling their fish to well-off foreigners, fish that they may never be able to afford for themselves.

Issues of "affluence guilt" aside,[1] it may be that farmers in the developing world benefit more from commercial, export-focused aquaculture than subsistence-based aquaculture. Subsistence aquaculture, by definition, does not provide farmers with more fish than

---

[1] Feelings of guilt over affluence are rooted in exposure to those less fortunate and the realization that their impoverished circumstances have been unduly influenced by their race, country, or other caprices of birth (Hoffman, 2000). Most consumers in the developed world are unaware of who catches or raises the seafood they eat, so perhaps their experiences are untarnished by thoughts of those in the developing world who raise their shrimp or salmon, but cannot afford to eat it.

their family can consume. They can achieve better nutrition and avoid the many indirect consequences of hunger or food instability,[2] but little else. Poor farmers engaged in commercial aquaculture and targeting export markets may not be able to afford to eat the fish they raise, but may be able to purchase more of other types of food. What is more, commercial aquaculture creates a value chain (including processing, marketing, and so forth) and supports other industries (feed manufacturing, transportation and distribution, refrigeration) creating economic opportunities beyond the farm (Subasinghe et al., 2012). In fact, it is estimated that for every one person directly employed in aquaculture, four others are indirectly supported by jobs in related industries (Food and Agriculture Organization, 2008). Medium- and larger-scale aquaculture enterprises can also provide workers with higher wages and other types of compensation, such as bonuses, healthcare, and retirement (Food and Agriculture Organization, 2016k).

It is somewhat counterintuitive, but raising high-value, carnivorous fish for export may help those in the developing world achieve food security more readily than raising herbivorous or omnivorous fish for direct, domestic consumption. That said, trophic level is not the only variable that influences product value, and rearing of low-value species at sufficient volume can still be highly profitable (Neori and Nobre, 2012). Tilapia and pangasid catfish illustrate this well: though they feed relatively low on the food chain, both are in high demand throughout most of the developed world and export markets for these fish can be very lucrative. Profitability is, of course, a function of product value and product cost. Intensive aquaculture systems are more expensive to operate and high-value species are more costly to raise, particularly in places where infrastructure may be absent or inadequate. Also, it is important to recognize that growth of intensive aquaculture can also hurt communities in the developing

---

[2] Malnutrition, particularly during childhood, weakens the immune system, interferes with cognitive development, and results in shorter lives with fewer opportunities (Food and Agriculture Organization, 2017n).

world if it is allowed to abuse the "commons" (see Chapter 19) and foster greater insecurity among the most vulnerable.[3]

In response to dwindling catches of the most valuable fish, capture fisheries have shifted from the largest, highest value carnivores to smaller species occupying lower trophic levels – a trend known as "fishing down the food web"; in aquaculture, the opposite trend – "farming up" – has been observed (Pauly et al., 2001). Aquaculture continues to be dominated by cultivation of herbivores and omnivores: these low-trophic level species represent nearly three-quarters of global aquaculture production.[4] The industry is nonetheless responding to the developed world's desire for carnivorous fish. Although the relative proportions of low, medium, and high trophic levels are not as imbalanced for aquaculture as for capture fisheries, aquaculture production is becoming increasingly top-heavy with high-value carnivores (Tacon et al., 2010). Some have argued that this is an unsustainable direction and that few of these high-value, high trophic level species should be raised, regardless of who can afford to consume them. "'Farming down the food web' [for example, raising filter feeders] has an environmentally positive meaning, while the opposite 'farming up the food web' [that is, raising top-level predators] is believed by some to be a more environmentally unsustainable and/or ethically illogical direction" (Tacon et al., 2010: 98). Most of this concern is related to the use of marine fisheries resources to feed farmed carnivorous fish instead of people (Naylor et al., 2009; Tacon and Metian, 2009), a topic dealt with elsewhere in this title (see Chapter 14).

In 1995, more than 170 member countries adopted the Food and Agriculture Organization *Code of Conduct for Responsible Fisheries*. Article 9 of the *Code* addresses aquaculture and includes provisions

---

[3] People who cannot afford other sources of animal protein are the most affected by the redirection of bycatch and "trash" fish to feeding higher value species in aquaculture. Impoverished people are also more likely to be impacted by pollution and environmental degradation, as they lack the mobility and other resources to avoid exposure (Subasinghe et al., 2012).

[4] The ratio is flipped for capture fisheries landings, with close to three-quarters being mid-level to top carnivores.

regarding ecological *and* economic sustainability, emphasizing judicious use of fisheries resources for aquaculture and other purposes, food security, and the needs of local communities (Tacon et al., 2010). Although a useful framework, it is difficult to determine how best to address the latter elements and the need for food versus economic opportunity. Teach a man to fish – or how to raise a fish – and you can feed him for a lifetime, but whether he chooses to eat his fish or sell them is another matter.

Chapter 19

Regulation of the aquaculture industry

The weightiest mistake in my synthesizing paper was the omission of the modifying adjective "unmanaged." (human ecologist, Garrett Hardin, reflecting on his most noteworthy work, "The Tragedy of the Commons," 1998: 683)

The phrase, "tragedy of the commons", is part of the scientific vernacular referring to the misuse and overexploitation of common property resources. It first appeared in an 1833 pamphlet by William Forster Lloyd describing the misuse of public resources by the public, using a hypothetical example of overgrazing of common property pastoral lands, that is, the "commons" (Lloyd, 1833). Cattle herders acting in their own personal interest and grazing more than their allotted number of livestock benefit in the short term, Lloyd argued, but the community suffers collectively in the long term as the commons become overgrazed and of decreasing value to all. Lloyd's phrase and economic theory were expanded and popularized by Garrett Hardin in his manuscript, "The Tragedy of the Commons" (Hardin, 1968). Following publication of Hardin's seminal work, the "commons" came to be recognized as a metaphor for natural resources such as freshwater, forests, fish stocks, the atmosphere, and so on. Twenty-five years after publication of "Tragedy", Hardin lamented that his most influential work had often been misinterpreted. He did not mean to suggest that exploitation of the world's natural resources is inherently "tragic" and to be avoided, but that unfettered access to a resource and complete freedom in pursuing our individual economic interests will lead to resource collapse and economic losses for both individuals and groups. Hardin said that such failures could – and should – be prevented by curtailing access or otherwise policing exploitation of the commons, adding that "The Tragedy of the Unmanaged or Unregulated Commons" might have been a better title (Hardin, 1998).

Regulation is essential to preventing degradation of the commons, whether it be a harvestable natural resource like timber or groundwater or a more nebulous resource such as functional ecosystems and the services they provide.[1] In the absence of regulation to prohibit or control man's activities, tragedies of the commons and other tragic consequences are likely.[2] Aquaculture, like other industrial activities relies on the commons – for water and land, feed resources and other raw material inputs, and energy – and must therefore be subject to regulation to avoid overexploitation of these

---

[1] The term "ecosystem services" was explicitly defined in the Millennium Ecosystem Assessment report (World Resources Institute, 2005) quite simply as the benefits that people receive from ecosystems. These include supporting services (photosynthesis, nutrient cycles, and other processes that allow ecosystems to function and life to persist), provisioning services (food, water, energy, and raw materials), regulating services (processes that establish or influence climate and weather patterns, water), and cultural services (recreational or esthetic values provided by the natural environment).

[2] Lack of appropriate safety and environmental protection laws and regulations have been associated with a wide range of environmental disasters. Notable examples include the Bhopal Disaster (chemical explosion and release of 40 tons of methyl isocyanate gas from a pesticide plant in Bhopal, India, in 1984, causing thousands of deaths and hundreds of thousands of injuries), Love Canal (improper use of a manmade canal for municipal and industrial waste disposal, exposing the community of Niagara Falls, USA, between 1978 and 2004, to toxic chemicals when the site and surrounding land were subsequently developed as commercial and residential property), and the Great Smog (London's notoriously poor air quality reached a nadir in the winter of 1952 when cold, windless weather and coal-related air pollution conspired to create a four-day smog event resulting in thousands of deaths and tens of thousands of respiratory illnesses). Each of these events precipitated the enactment of major federal laws to protect workers and prevent further industrial abuses of the commons.

shared resources. The fact that aquaculture is, indeed, regulated to prevent misuse of the commons does not seem to be widely understood or appreciated: "Unregulated aquaculture" is the subject of a variety of online videos, blog posts, popular and technical articles. Collectively, these sources suggest that aquaculture operates in a regulatory vacuum – a sort of lawless "high seas" of seafood. In truth, aquaculture is regulated – to varying degrees – based on what aquatic organism is being raised and where.

Regulation of aquaculture varies among and within countries, but whatever oversight is in place usually addresses a number of common elements intended to ensure industry pays the full cost of producing farmed seafood. For example, regulations prohibiting the discharge of nutrient-laden effluents from fish farms are intended to force farmers to pay for filtration systems or some other means of cleaning up their effluents. In other words, these regulations make sure that farmers do not "pass the buck" for waste management and abuse the water purifying ecosystem services of the receiving waters.[3] Aquaculture regulations typically include oversight of land and water use; wastewater discharges; drug and chemical use; animal welfare and food safety standards; control of fish movements, pathogens, and invasive species; and an overall oversight and enforcement scheme by which the various activities are permitted and monitored (Hishamunda et al., 2014) (Figure 19.1).

As is often the case with environmental regulation, many aquaculture laws and policies have been promulgated reactively in response to observations of declining environmental quality, either as a result of aquaculture or other anthropogenic influences.[4]

---

[3] For fans of economics, this is known as "internalizing the externalities." Aquaculture is associated with a number of externalities or side effects, depending on the type of aquaculture. Most are negative – pollution, habitat loss, effects on wild fish populations – but some types of aquaculture generate positive externalities, such as oyster farming improving water clarity. Punitive regulations are the primary means by which industries are forced to internalize the unintended consequences of their activity, but policies can also incentivize industries that create positive externalities.

[4] For instance, the U.S. Clean Water Act (CWA) was developed primarily to control discharges of chemical waste and sewage into the country's surface waters,

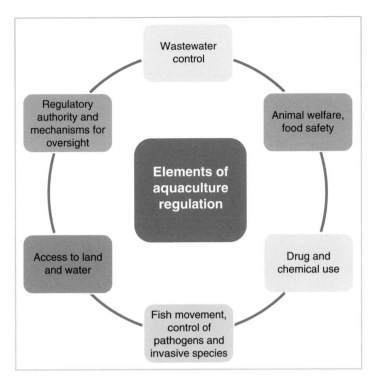

Figure 19.1: Regulatory oversight of aquaculture varies from place to place and can differ based on the cultured species and methods (for example, finfish aquaculture is often regulated in a different manner than cultivation of filter-feeding bivalves). That said, regulations typically address a number of elements intended to curtail environmental effects and limit consequences for the "commons", and involve one or more authorities responsible for creating a permitting and monitoring scheme and enforcing the rules (Hishamunda et al., 2014).

Freshwater eutrophication has been associated with aquaculture in North and South America, Asia, and Europe; nonnative fish species have escaped farms on every continent; overuse of drugs and other chemicals has been a problem in Asia and South America, and so on (Zhu and Chu, 2013). To be considered legitimate by the public, any mechanism of governmental oversight should meet standards of accountability (government is responsive to the needs of its citizens),

but wastewater generated by large fish farms is also subject to the CWA and its associated regulations (U.S. Environmental Protection Agency, 2017a).

participation (those participating in and affecting by the regulated activity help shape laws and regulations), predictability (laws are applied fairly and consistently and do not change except according to established processes), and transparency (decision-making processes are open and clear), and be executed by an authority with the capacity and competence to ensure implementation adheres to these standards. Governance of aquaculture is no different – it should be built on a foundation of legitimacy – and should ideally enable industry growth by providing infrastructure, research, and education support and help the industry to connect with citizens to foster public "buy in" (Hishamunda et al., 2014).

Different nations fulfill these governance goals to varying degrees (Figure 19.2) (Zhu and Chu, 2013). Aquaculture operations should be sited properly, taking the carrying capacity of the surrounding ecosystem into account – Australia and Norway have such programs in place. Licenses or permits should include reasonable time limits after which the licensee must reapply or return the licensed area to its original condition, as is required in Norway. Monitoring is essential to ensure fish farmers are staying within the limits established by their licenses or permits – Australia, Norway, and the United States all have such comprehensive environmental monitoring programs in place. Of course, such monitoring is most effective when there are standards in place that define what effluent volumes and pollutant discharges are considered acceptable, as is the case in the United States, Australia, and most European countries. Australia and Thailand take the process of waste management a step further by stipulating certain aspects of production processes, feed formulations, and so on, that help to minimize discharges. Drug and chemical usage is restricted by most countries, though the list of allowable treatments varies from place to place (see Chapter 7).

Regulation is essential, particularly as our world becomes more populous and resources become scarcer, but regulation can curb growth. Failing to provide regulatory protections for workers and the environment can have dire consequences, but industries are often attracted to locations where the costs of regulatory compliance

Figure 19.2: Although environmental regulation, including oversight of aquaculture, is generally lacking in developing countries, most developed countries have enacted laws to comprehensively address the environmental impacts of raising fish and shellfish. The United States, Australia, Norway, and South Korea closely regulate aquaculture in all major oversight categories; Thailand and China also exert considerable oversight of aquaculture within their borders, but do not require treatment of effluent or environmental impact assessments of aquaculture operations (Zhu and Chu, 2013).

are lower.[5] The same appears to be true for aquaculture, in that more stringent environmental regulations have been associated with slower growth or contraction of the aquaculture industry (Abate et al., 2016) (Figure 19.3).

Beyond the black and white of what is allowed and what is not, there are a variety of ways in which laws, regulations, or policies affect aquaculture (Knapp and Rubino, 2016). For example, the nature of the permitting process – its predictability, inherent costs, flexibility,

---

[5] Either because the regulations do not exist, are "lax", or are not enforced.

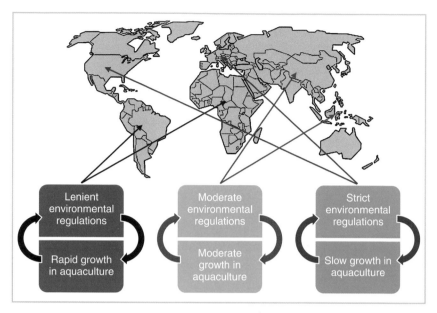

Figure 19.3: The relative stringency by which countries regulate their aquaculture industries appears to influence the speed at which they have grown since the dawning of the "blue revolution" (Abate et al., 2016). Regions with more lenient environmental regulations tend to have faster growing aquaculture sectors (Africa and Latin America, 11.2–12.3% growth during 1990s and 2000s) than those with more moderate (Asia and Oceania, 7.7–9.4% growth) or strict environmental regulations (North America and Europe, 1.5–4.5% growth).

Source: World map image / Author: Tom-b / Wikimedia Commons / Public Domain.

and resilience to legal challenges – can influence how many aquaculture operations are permitted, become established, and stay in business as much as the requirements of the permits themselves. Costs of compliance and predictability also determine how environmental regulations influence aquaculture businesses. Governmental policies such as taxation and subsidies or other incentives also have an effect on the viability and growth of aquaculture, as do the presence and relative strength of support programs (research, development, and extension) and infrastructure.

This is not to say that we should model regulatory oversight of aquaculture based on the practices of some developing nations in the hopes of matching the growth of their fish farming sectors. The tragedy of the commons teaches us that the short-term gains extracted from unmanaged commons will not last. Aquaculture conducted in an environmentally exploitative manner will ultimately falter (such as salmon farming in Chile or turbot farming in China), but aquaculture need not be throttled into near nonexistence by unnecessarily burdensome regulation (as is the case for salmon farming in the United States and Ireland) (Osmundsen et al., 2017). Instead, the goal should be regulatory efficiency and policies that enable aquaculture practices that do not "overgraze" the commons. Unfortunately, achieving regulatory efficiency is easier said than done: governance of aquaculture is a "wicked problem", characterized by uncertainty, lack of consensus, and an ever-changing field of play (Osmundsen et al., 2017).

The United States provides a particularly useful example of how regulatory complexity and unpredictability can constrain an industry. At roughly 425,000 metric tons per year, the United States is the 18th largest producer of farmed fish and shellfish (Food and Agriculture Organization, 2018g). The United States' aquaculture industry is dwarfed by that of other nations with less abundant freshwater resources, shorter coastlines, and substantially less demand for seafood. Given all that the United States has working in its favor, why is the American aquaculture industry being outstripped by that of Myanmar, Ecuador, and 15 other nations? The answer may lie in its regulatory climate: United States aquaculture is subject to more than 1300 laws governing environmental management, food safety, legal and labor standards, interstate transport of aquatic products, fish health, and culture of species that are also commercially fished (Engle and Stone, 2013). The aquaculture industry should be answerable to authorities charged with protecting workers, the environment, and consumers, but one wonders whether the United States' approach is effective or efficient in that regard. What is clear is that the cost of regulatory compliance is an albatross for American fish and shellfish farmers: permit fees, fish health testing, lost sales, and staff time spent dealing with regulatory matters may add up to nearly US$150,000 annually per farm (van Senten and Engle, 2017).

The Norwegians have adopted a different approach to aquaculture regulation. Like the United States, Norway exerts considerable regulatory control of aquaculture: on a scale of 1 (very lax) to 7 (among the most stringent in the world), U.S. environmental regulations ranked an impressive 5.2, Norwegian regulations even higher at 5.8 (Abate et al., 2016). But as the eighth largest aquaculture producer in the world, Norway produces more than three times as much farmed seafood as the United States (Food and Agriculture Organization, 2018g). Norway has been able to sustain a rapidly growing aquaculture industry while continuing to enforce rigorous environmental controls principally through regulatory efficiency. Whereas fish farmers in the United States must engage with dozens of municipal, regional, state, and federal authorities to fully satisfy regulatory requirements, Norwegian fish farmers enjoy a process akin to "one-stop shopping": everything they need can be found in essentially one place (Engle and Stone, 2013).

Various factors including regulatory climate determine whether aquaculture is a vibrant, growing enterprise or a shrinking, misunderstood industry, but the Norwegian example suggests that aquaculture can be simultaneously fostered and kept to high environmental standards.

PART V

Conclusions

Chapter 20

Separating fact from fiction and advocating for aquaculture

Everyone is entitled to his own opinion, but not to his own facts. (Daniel Patrick Moynihan, American ambassador, senator, and presidential advisor)

Start where you are. Use what you have. Do what you can. (Arthur Ashe, American tennis legend and HIV/AIDS activist)

Moynihan's wry, pithy defense of truth begins *Fantasyland*, Kurt Andersen's treatise on the increasing disconnection between American life and objective reality (Andersen, 2017a). In it, he explores the concepts of "post-truth" and "post-factual" and how these nonsensical terms have gained meaning and come to define how we view ourselves and our world. Both can trace their origins to a common linguistic forebear, the seemingly harmless "truthiness." Coined by Stephen Colbert, truthiness is "the belief in what you feel to be true rather than what the facts will support." Colbert's satirical word-smithing captured the then-nascent distrust among American political factions and the emergence of traditional and social media that made it possible for the public to consume only that which reinforced what they already believed. Truthiness was recognized as the 2006 "Word of the Year" by Merriam-Webster Dictionary; ten years later, post-truth was similarly honored by Oxford Dictionaries. In the post-truth world, "objective facts are less influential in shaping public opinion than appeals to emotion and personal belief" (Keane, 2016).

Although the distinction between facts and opinions may be considered passé in the post-truth era, Moynihan's quote also begins this chapter, the last in *Understanding Aquaculture*. It appears above to remind those reading these pages that any debate over aquaculture must be grounded in objectivity. Aquaculture is a misunderstood and maligned enterprise, tarred with claims that it damages the environment, consumes more fish than it produces, and delivers poor quality food that is unsafe and less nutritious than what we could simply harvest from the sea. There is precious little truth to these "truthy" claims. The amount of seafood produced by the aquaculture industry vastly outweighs the volume of wild-caught fish that are used as raw materials for feed manufacturing. The nutritional value of farmed fish and shellfish is equivalent, sometimes even superior to that of wild-caught seafood. Contaminants have been detected in both wild and farmed fish, but the balance of evidence suggests that farmed seafood is generally safer to eat. Aquaculture does have an environmental footprint – one cannot produce nearly 74 million metric tons of anything without using inputs and creating outputs[1] – but striking improvements have been made to make aquaculture more environmentally sustainable and considerably less resource-consumptive than many forms of animal agriculture. Perhaps the most dangerous of all these post-factual allusions is that aquaculture is not needed, that we can feed ourselves now and in the future by just catching more fish. The truth – in the traditional, objective, fact-based sense – is that aquaculture is a necessity. Without it, we cannot hope to feed ourselves without decimating capture fisheries and missing the opportunity to use other natural resources more judiciously.

Aquaculture has a good story to tell, and the facts are on our side. Why is the public so willing to accept the many falsehoods about fish farming as facts? Why is the less-than-accurate, subjective narrative about aquaculture so much more effectively communicated and embraced than the objective truth?

---

[1] For those interested in a more tangible representation of annual global aquaculture production, 74 million metric tons is more than the weight of ninety Golden Gate Bridges (~800,000 metric tons) or one million space shuttles (~70 metric tons).

Certain ideas are infectious, and once they start to spread, they become increasingly difficult to contain and eradicate. We use the vernacular of diseases and epidemics to describe ideas that take over public discourse or pop culture – things that "go viral". But what makes an idea infectious? What made, for example, the ice bucket challenge different from any other campaign to raise awareness about ALS[2] (The ALS Association, 2018)? These are questions that were explored by Malcolm Gladwell nearly 20 years ago in his seminal book, *The Tipping Point* (Gladwell, 2000). Recognizing that ideas, products, messages, and behaviors can be contagious, what is it that makes some of them truly viral? Why is it, for instance, that everyone from school children to seniors seems to think that fish have a 3-second memory? If one ponders the idea – even for just a few seconds – it becomes clear that this cannot possibly be true, not really. If fish could not remember anything, how could they learn to recognize and avoid the signs of predators in the environment? Some might argue that predator avoidance is instinctually hard-wired, not learned behavior. Maybe. But what about the way a pet goldfish swims right up to the glass at the sight of its owner, eagerly gulping for the flakes of food before they are even offered. If they could not remember anything for more than a few seconds, goldfish would not be able to recall the meals that came shortly after a lumbering humanoid presence approached their bowl. The truth is that fish are possessed of cognitive functions far beyond simple learned behavior, engage in a wide range of complex social and strategic behaviors, and have well-developed short- and long-term memories (Laland et al., 2003). The idea of fish having a 3-second memory is demonstrably false, but it is persistent and pervasive; to use Gladwell's parlance, it is an idea that has tipped.

One of the most important attributes of an infectious idea is something Gladwell called the "stickiness factor." The nature of a message – both the message itself and the way in which it is presented – can determine whether it is memorable or forgettable. One of the most important attributes of sticky ideas is simplicity. Simple ideas are

---

[2] Amyotrophic lateral sclerosis, also known as Lou Gehrig's disease.

easier to understand and remember – they stick in people's minds more than complex ideas. It is something of a running joke among scientists that the answer to any question is, "It depends." It is funny because it is true. If you ever find yourself at a professional conference or some other gathering of scientific minds, try it out. Ask a question and you are likely to get some equivocation followed by a lengthy, detailed explanation of how the answer varies from one scenario to another. There is very little in science that is absolute, and the answers are rarely black and white. Scientists are not wrong for communicating in the fuzzy, gray language of "it depends" because the specific circumstances really do matter. The true nature of memory in fish is perhaps best, most fully understood by those immersed in the gray, who can draw meaning from the low-contrast differences among species, experimental designs, and ways of measuring memory. For those who have read the many hundreds of peer-reviewed research articles on the subject and can appreciate the nuance, it is possible to conjure some understanding of fish memory from the gray-shaded contours. As more experiments are done and the layers of gray accumulate, the picture may become clearer, but it will never be completely black and white. The answer will still be, "It depends." The myth that fish have a 3-second memory wastes no time on mealy mouthed hedgewords or shades of gray – it is black and white, simple and truthy. To many people, fish seem like they probably would have short memories, so they believe, remember, and tell others that they do. Complex ideas – about fish memory or the essential role of aquaculture – are not very likely to be infectious. They simply are not as sticky or seemingly right as simple, truthy ideas.

So where does this leave us? The truth about aquaculture is complicated – a Rube Goldberg[3]-esque assemblage of facts and figures that most people are unlikely to seek out, if they are interested at all. For those of us who do understand aquaculture and hope to see

---

[3] Reuben Garrett Lucius Goldberg was an American cultural multihyphenate best known for his cartoons depicting comically complicated machines that were purported to save time and simplify life, but did neither in practice. As the popularity of Goldberg's works grew, his name became synonymous with unnecessarily complex gadgetry.

it fulfill its promise, how do we counter the waves of misinformation that wash over the public consciousness and fight the rising tides of do-it-yourself, post-truth reality? Unfortunately, there is no easy solution or twelve-step program to guide us back to the once-traditional norms of truth and objectivity. But we are not powerless. In his final thoughts on the way forward in a post-factual fantasyland, Andersen offered a few, mostly self-reliant recommendations:

> What is to be done? I don't have an actionable agenda … But I think we can slow the flood, repair the levees, and maybe stop things from getting any worse … We need to firmly commit to Moynihan's aphorism about opinions versus facts. We must call out the dangerously untrue and unreal … Fight the good fight in your private life. You needn't get into an argument with [a] stranger, … but do not give acquaintances and friends and family members free passes … And fight the good fight in the public sphere … Progress is not inevitable, but it's not impossible, either. (Andersen, 2017b)

If you have read these pages, you understand aquaculture in a way that many do not. You are prepared to engage thoughtfully the next time aquaculture comes up in conversation and to fight the good fight. You might even be motivated to bring the subject up yourself – I do every time I order seafood at a restaurant. Before I make my selection, I find out whether the fish on the menu is farmed or wild. Most waiting staff are not taken aback by this question – these days, it is fairly commonplace for seafood origin to be clearly indicated on the menu – but most are surprised when I insist on ordering only farmed fish or shellfish. I explain that freshness, taste, value, and sustainability matter to me, and that farm-raised seafood is good, good for me, and good for the environment. I believe in aquaculture, and whether it is at a restaurant or the grocery store, I put my money where my mouth is – I buy farm-raised seafood. Out of loyalty or obligation, many of my friends and family have given farm-raised seafood a try. Having tried it themselves, even the most skeptical are now willing to admit that farmed does not mean inferior. One friend spent the first half of his career in Alaska, raising and releasing millions of juvenile salmon that would ultimately return from the sea to be caught and sold as wild fish. Several years ago, he sheepishly admitted to having tried sushi prepared with farm-raised Atlantic salmon. He was a little embarrassed, not for having tried it, but for choosing farmed salmon over wild again and again because it was "just that good." I doubt my friend would ever be caught dead publicly extolling the virtues of farm-raised seafood, but he no longer bemoans aquaculture products as a second-rate. And I know he is quietly supporting aquaculture every time he visits a sushi restaurant.

Agriculture was the harbinger of civilization, irreversibly changing the trajectory of human evolution. Aquaculture, too, is transforming the way we live. Aquaculture is providing essential nutrition and economic opportunity where it is needed most in the world. Aquaculture is helping current and future generations to be healthier and better fed, without razing aquatic or terrestrial ecosystems. For those lucky enough to not worry about food security, aquaculture still offers the comfort and satisfaction of a well-cooked meal, as wholesome as it is delicious. Aquaculture supports fisheries that would otherwise collapse under the influences of fishing pressure and habitat loss and pulls threatened and endangered fish back from the brink of extinction. Aquaculture provides flamboyantly colored fish for our enjoyment as pets as well as hardy aquatic guinea pigs needed for biomedical testing and other research. Aquaculture feeds us, provides us with jobs and entertainment, answers important scientific questions, and helps to protect the natural and cultural legacy we will give future generations.

Fifty years ago, aquaculture was little more than a novelty, insignificant by nearly all measures; today it is the fastest growing form of agriculture and the most important source of seafood in the world. Aquaculture is undoubtedly ascendant, but its full potential has not yet been realized. If we are to fully reap the rewards of the blue revolution, we must acknowledge the need for aquaculture, address its biological and social pitfalls, and foster its continued growth. *Start where you are.* Visit a local fish farm or hatchery and make farm-raised fish or shellfish your next meal. *Use what you have.* Support aquaculture with your dollars, your voice, and your votes. *Do what you can.* Understand aquaculture, embrace it, and help others to do the same.

Taxonomic glossary of common and scientific names

Abalones, *Haliotis* spp.

Alaskan pollock or walleye pollack, *Gadus chalcogrammus*

Argentine shortfin squid, *Illex argentines*

Asian carps, various cyprinid species native to Asia, excluding those native to the Indian subcontinent

Atlantic cod, *Gadus morhua*

Atlantic salmon, *Salmo salar*

Barramundi, *Lates calcarifer*

Bighead carp, *Hypophthalmichthys nobilis*

Black pacu, *Colossoma macropomum*

Black spot sea bream, *Pagellus bogaraveo*

Black tetra *Gymnocorymbus ternetzi*

Blue crabs, *Portunus* and *Callinectes* spp.

Blue mussel, *Mytilus edulis*

Bluefin tunas, various *Thunnus* spp.

Brook trout, *Salvelinus fontinalis*

Brown trout, *Salmo trutta*

Catla, *Catla catla*

Channel catfish, *Ictalurus punctatus*

Chinook salmon, *Oncorhynchus tshawytscha*

Chum salmon, *Oncorhynchus keta*

Cobia, *Rachycentron canadum*

Coho salmon, *Oncorhynchus kisutch*

Common carp, *Cyprinus carpio*

Cutthroat trout, *Oncorhynchus clarkia*

Flathead grey mullet, *Mugil cephalus*

Giant river prawn, *Macrobrachium rosenbergii*

Giant tiger prawn, *Penaeus monodon*

Gilthead sea bream, *Sparus aurata*

Golden dorado, *Salminus brasiliensis*

Goldfish, *Carassius auratus*

Grass carp, *Ctenopharyngodon idella*

Green mussel, *Perna viridis*

Groupers, *Epinephelus* spp. and *Cromileptes* spp.

Indian major carps, collective term for catla, rohu, and mrigal

Japanese carpet shell or Manila clam, *Venerupis philippinarum*

Jumbo flying squid or Humboldt squid, *Dosidicus gigas*

Katoni seaweed, *Kappaphycus alvarezii*

Korean mussel, *Mytilus coruscus*

Lahontan cutthroat trout, *Oncorhynchus clarki henshawi*

Lake trout, *Salvelinus namaycush*

Lobsters, *Panulirus* spp.

Maculated ivory whelk, *Babylonia micropeltes*

Mahi-mahi or common dolphin-fish, *Coryphaena hippurus*

Milkfish, *Chanos chanos*

Mola carplet, *Amblypharyngodon mola*

Mrigal, *Cirrhinus mrigala*

Mullets, *Mugil* spp.

Nile tilapia, *Oreochromis niloticus*

Ocean pout, *Zoarces americanus*

Pacific cupped oyster, *Crassostrea gigas*

Pacific whiteleg shrimp, *Litopenaeus vannamei*

Pearl oysters, *Pinctada* spp.

Penaeid shrimps, *Penaeus* and *Litopenaeus* spp.

Pink salmon, *Oncorhynchus gorbuscha*

Rainbow trout or steelhead, *Oncorhynchus mykiss*

Rohu, *Labeo rohita*

Sauger, *Sander canadense*

Sea bass, *Dicentrarchus labrax*

Siberian sturgeon, *Acipenser baerii*

Silver carp, *Hypophthalmichthys molitrix*

Skipjack tuna, *Katsuwonus pelamis*

Slipper cupped oyster, *Crassostrea iredalei*

Snakeheads, *Channa* spp.

Snow or spider crabs, *Chinoecetes* spp.

Sockeye salmon, *Oncorhynchus nerka*

Striped bass, *Morone saxatilis*

Tiger barb, *Puntius tetrazona*

Tilapias, various genera of tilapiine cichlids

Turbot, *Scophthalmus maximus*

Vietnamese catfishes, *Pangasius* spp.

Walleye, *Sander vitreus*

White bass, *Morone chrysops*

White perch, *Morone Americana*

Yellow bass, *Morone mississippiensis*

Yellow perch, *Perca flavescens*

Yellowfin tuna, *Thunnus albacares*

Zebra danio, *Danio rerio*

Works cited

Aarste, B., Beckmann, S., Bigne, E., Beveridge, M., Bjorndal, T., Bunting, J., McDonagh, P., Mariojouls, C., Muir, J., Prothero, A., Reisch, L., Smith, A., Tveteras, R., Young, J. (2004). The European consumers' understanding and perceptions of the "organic" food regime: the case of aquaculture. British Food Journal, 106, 93–105.

Abate, T., Neilsen, R., and Tveteras, R. (2016). Stringency of environmental regulation and aquaculture growth: a cross-country analysis. Aquaculture Economics and Management, 20, 201–221.

Aguilar-Manjarrez, J., Soto, D., and Brummett, R. (2017). Aquaculture zoning, site selection, and area management under the ecosystem approach to aquaculture, a handbook. Rome, Italy: Food and Agriculture Organization of the United Nations and the World Bank.

Amberg, S., and Hall, T. (2008). Communicating risks and benefits of aquaculture: a content analysis of US newsprint representations of farmed salmon. Journal of the World Aquaculture Society, 39, 143–157.

American Pet Products Association. (2016). 2015–2016 APPA national pet owners survey. Greenwich, CT: American Pet Products Association.

Amos, K., Gustafson, L., Warg, J., Whaley, J., Purcell, M., Rolland, J., Winton, J.R., Snekvik, K., Meyers, T., Stewart, B., Kerwin, J., Blair, M., Bader, J., and Evered, J. (2014). U.S. Response to a report of infectious salmon anemia virus in western North America. Fisheries, 39, 501–506.

Anderson, J. (2007). Aquaculture and fisheries: complement or competition. In R. Arthur, and J. Neirentz, Global Trade Conference on Aquaculture (pp. 231–236). Rome, Italy: Food and Agriculture Organization of the United Nations.

Andersen, K. (2017a, September). Fantasyland: How America Went Haywire, A 500-year History. New York: Random House.

Andersen, K. (2017b, September). How America lost its mind. The Atlantic. Retrieved March 13, 2018, from https://www.theatlantic.com/magazine/archive/2017/09/how-america-lost-its-mind/534231/

Anon. (1988, September 11). On language: rot at the top. New York Times Magazine. Retrieved August 27, 2017, from www.nytimes.com/1988/09/11/magazine/on-language-rot-at-the-top.html?mcubz=1

Anon. (2003, April 14). Sabotage suspected in mass fish getaway. ABC News Australia. Retrieved September 10, 2017, from www.abc.net.au/news/2003–04–15/sabotage-suspected-in-mass-fish-getaway/1836764

Anon. (2005, November 25). Fish lost in vandalism valued at $2.5M. TheFishSite. Retrieved September 10, 2017, from https://thefishsite.com/articles/fish-lost-in-vandalism-valued-at-25m

Aranceta, J., and Perez-Rodrigo, C. (2012). Recommended dietary reference intakes, nutritional goals and dietary guidelines for fat and fatty acids: a systematic review. British Journal of Nutrition, 107, S8–S22.

Arechavala-Lopez, P., Uglem, I., Fernandez-Jover, D., Bayle-Sempere, J., and Sanchez-Jerez, P. (2011). Immediate post-escape behaviour of farmed seabass (*Dicentrarchus labrax L.*) in the Mediterranean Sea. Journal of Applied Ichthyology, 27, 1375–1378.

Arechavala-Lopez, P., Sanchez-Jerez, P., Bayle-Sempere, J., Uglem, I., and Mladineo, I. (2013). Reared fish, farmed escapees and wild fish stocks–a triangle of pathogen transmission of concern to Mediterranean aquaculture management. Aquaculture Environment Interactions, 3, 153–161.

Arismendi, I., Soto, D., Penaluna, B., Jara, C., Leal, C., and and Leon-Munoz, J. (2009). Aquaculture, non-native salmonid invasions and associated declines of native fishes in Northern Patagonian lakes. Freshwater Biology, 54, 1135–1147.

Arthur, R., Lorenzen, K., Homekingkeo, P., Sidavong, K., Sengvilaikham, B., and Garaway, C. (2010). Assessing impacts on introduced aquaculture species on native fish communities: Nile tilapia and major carps in SE Asian freshwaters. Aquaculture, 299, 81–88.

Arvanitoyannis, I., Krystallis, A., Panagiotaki, P., and Theodorou, A. (2004). A marketing survey on Greek consumers' attitudes toward fish. Aquaculture International, 12, 259–279.

Asche, F., and Tveteras, S. (2004). On the relationship between aquaculture and reduction fisheries. Journal of Agricultural Economics, 55, 245–265.

Asche, F., Guttormsen, A., Sebulonsen, T., and Sissener, E. (2003). Competition between farmed and wild salmon: the Japanese salmon market, working paper number 44/03. Bergen, Germany: Institute for Research in Economics and Business Administration.

Ashton, E. (2008). The impact of shrimp farming on mangrove ecosystems. CAB Reviews: Perspectives in Agriculture, Veterinary Science, Nutrition and National Resources, 3, 1–12.

Bacher, K. (2015). Perceptions and misconceptions of aquaculture: a global overview, Globefish Research Program Volume 120. Rome: Food and Agriculture Organization of the United Nations.

Bailey, C. (2015). Transgenic salmon: science, politics, and flawed policy. Society and Natural Resources, 28, 1249–1260.

Barney, G. (1980). Global 2000 Report to the President of the U.S.: Entering the 21st Century. New York: Pergamon Press.

Barrett, K., Nakao, J., Taylor, E., Eggers, C., and Gould, L. (2017). Fish-associated foodborne disease outbreaks: United States, 1998–2015. Foodborne Pathogens and Disease. doi:https://doi.org/10.1089/fpd. 2017.2286

Bartholomew, J., and Reno, P. (2002). The history and dissemination of whirling disease. In J. Bartholomew, and C. Wilson, Whirling Disease: Reviews and Current Topics. Bethesda, MD: American Fisheries Society.

Baskett, M., Burgess, S., and Waples, R. (2013). Assessing strategies to minimize unintended fitness consequences of aquaculture on wild populations. Evolutionary Applications, 6, 1090–1108.

Bauer, J. (2006). Metabolic basis for the essential nature of fatty acids and the unique dietary fatty acid requirements of cats. Journal of the American Veterinary Medical Association, 229, 1729–1732.

Baz, G., Hanaa, F., Salwa, A.-F., and Saleh, O. (2014). Comparative study on spoilage markers and chemical composition of farmed and wild *Oreochromis niloticus*. Journal of the Arabian Aquaculture Society, 9, 171–182.

Bergh, Ø. (2007). The dual myths of the healthy wild fish and the unhealthy farmed fish. Diseases of Aquatic Organisms, 75, 159–164.

Berlanga-Robles, C., Ruiz-Luna, A., and Hernandez-Guzman, R. (2011). Impact of shrimp farming on mangrove forest and other coastal wetlands: the case of Mexico. In B. Sladonja, Aquaculture and the Environment, A Shared Destiny (pp. 17–30). Rijeka, Croatia: InTech.

Berntssen, M., Lundebye, A.-K., and Torstensen, B. (2005). Reducing the levels of dioxins and dioxin-like PCBs in farmed Atlantic salmon by substitution of fish oil and vegetable oil in the feed. Aquaculture Nutrition, 11, 219–231.

Blanchet, C., Lucas, M., Julien, P., Morin, R., Gingras, S., and Dewailly, E. (2005). Fatty acid composition of wild and farmed Atlantic salmon (*Salmo salar*) and rainbow trout (*Oncorhynchus mykiss*). Lipids, 40, 529–531.

Blanchfield, P., Tate, L., and Podemski, C. (2009). Survival and behavior of rainbow trout (*Oncorhynchus mykiss*) released from an experimental aquaculture operation. Canadian Journal of Fisheries and Aquatic Science, 66, 1976–1988.

Blaylock, R., and Bullard, S. (2014). Counter-insurgents of the blue revolution? Parasites and diseases affecting aquaculture and science. Journal of Parasitology, 100, 743–755.

Bocking, S. (2010). Mobile knowledge and the media: the movement of science information in the context of environmental controversy. Public Understanding of Science, 21, 705–723.

Borderias, A., and Sanchez-Alonso, I. (2011). First processing steps and the quality of wild and farmed fish. Journal of Food Science, 76, R1-R5.

Bosma, R., and Verdegem, M. (2011). Sustainable aquaculture in ponds: principles, practices, and limits. Livestock Science, 139, 58–68.

Boswell, E. (2009, April 27). Has whirling disease come full circle? Retrieved from Mountains & Minds, Montana State University, http://www.montana.edu/news/mountainsandminds/article.html?id=9115

Bowker, J., and Trushenski, J. (2015). Mythbusters: What's real and what's not when it comes to fish drugs. North American Journal of Aquaculture, 77, 358–366.

Boxaspen, K. (2006). A review of the biology and genetics of sea lice. ICES Journal of Marine Science, 63, 1304–1316.

Braun, A. (2017, August 28). Farmed and dangerous? Pacific salmon confront rogue Atlantic cousins. Retrieved January 27, 2018, from Scientific American: https://www.scientificamerican.com/article/farmed-and-dangerous-pacific-salmon-confront-rogue-atlantic-cousins/

Bronnmann, J., and Asche, F. (2017). Sustainable seafood from aquaculture and wild fisheries: insights from a discrete choice experiment in Germany. Ecological Economics, 142, 113–119.

Butler, D., and Reichhardt, T. (1999). Long-term effect of GM crops serves up food for thought. Nature, 398, 651–653.

Calow, P. (1998). The encyclopedia of ecology and environmental management. Oxford, UK: Blackwell Publishers.

Camden, J. (2018, January 12). Washington could end net pens for fish by 2024. The Spokesman-Review. Spokane, Washington. Retrieved from: http://www.spokesman.com/stories/2018/jan/12/washington-co uld-end-net-pens-for-fish-by-2024/

Canzi, C., Roubach, R., Benassi, S., Matos, F., Motter, I., and Bueno, G. (2017). Selection of sites for establishing aquaculture parks, and estimation of fish production carrying capacity for a tropical reservoir in South America. Lakes Reservoirs, 22, 148–160.

Cardoso, C., Lourenco, H., Costa, S., Goncalves, S., and Nunes, M. (2013). Survey into the seafood consumption preferences and patterns in the Portuguese population, gender and regional variability. Appetite, 64, 20–31.

Carrington, C., and Bolger, M. (n.d.). An exposure assessment for methylmercury from seafood for consumers in the United States. Washington, D.C.: U.S. Food and Drug Administration.

Centers for Disease Control and Prevention. (2015a, August 10). Heart Disease Facts. Retrieved March 22, 2017, from https://www.cdc.gov/heartdisease/facts.htm

Centers for Disease Control and Prevention. (2015b, May 27). New CDC data on foodborne disease outbreaks. Retrieved August 27, 2017, from CDC Features: https://www.cdc.gov/features/foodborne-diseases-data/index/html

CGIAR. (n.d.). Food Security. Retrieved July 4, 2017, from Big Facts: https://ccafs.cgiar.org/bigfacts/#theme=food-securityandsubtheme=diets

Chef's Resources. (n.d.). Finfish butchering yield % to fillets and loins. Retrieved July 10, 2016, from Chef's Resources: http://www.chefs-resources.com/seafood/seafood-yields/

Chiaramonte, L., Munson, D., and Trushenski, J. (2016). Climate change and considerations for fish health and fish health professionals. Fisheries, 41, 396–399.

Chittenden, C., Rikardsen, A., Skilbrei, O., Davidsen, J., Halttunen, I., Skardhamar, J., and McKinley, R. (2011). An effective method for the recapture of escaped farmed salmon. Aquaculture Environment Interactions, 1, 215–224.

Christensen, V., Coll, M., Piroddi, C., Steenbeek, J., Buszowski, J., and Pauly, D. (2014). A century of fish biomass decline in the ocean. Marine Ecology Progress Series, 512, 155–166.

Christie, W. (2011, April 1). Fatty acids and eicosanoids. Retrieved January 8, 2017, from AOCS Lipid Library: http://lipidlibrary.aocs.org/Primer/content.cfm?ItemNumber=39290&navItemNumber=19201

Christie, W. (2014, July). Resolvins and protectins. Retrieved January 8, 2017, from AOCS Lipid Library: http://lipidlibrary.aocs.org/Primer/content.cfm?ItemNumber=39317

Claret, A., Guerrero, L., Aguirre, E., Rincon, L., Hernandez, M., Martinez, I., Benito Peleteiro, J., Grau, A., Rodriguez-Rodriguez, C. (2012). Consumer preferences for sea fish using conjoint analysis: exploratory study of the importance of country of origin, obtaining method, storage conditions and purchasing price. Food Quality and Preference, 26, 259–266.

Claret, A., Guerrero, L., Gines, R., Grau, A., Hernandez, M., Aguirre, E., Benito, Peleteiro, J., Fernández-Pato, C., Rodriguez-Rodriguez, C. (2014). Consumer beliefs regarding farmed versus wild fish. Appetite, 79, 25–31.

Claret, A., Guerrero, L., Gartzia, I., Garcia-Quiroga, M., and Gines, R. (2016). Does information affect consumer liking of farmed and wild fish? Aquaculture, 454, 157–162.

Conte, F., Passantino, A., Longo, S., and Voslarova, E. (2014). Consumers' attitude towards fish meat. Italian Journal of Food Safety, 3, 178–181.

Cordain, L., Eaton, S., Sebastian, A., Mann, N., Lindeberg, S., Watkins, B., O'Keefe, J.H., Brand-Miller, J. (2005). Origins and evolution of the Western diet: health implications for the 21st century. American Journal of Clinical Nutrition, 81, 341–354.

Coutant, C. (1998). What is "Normative" for Fish Pathogens? A Perspective on the Controversy over Interactions between Wild and Cultured Fish. Journal of Aquatic Animal Health, 10, 101–106.

Crawford, M., Bloom, M., Leigh Broadhurst, C., Schmidt, W., Cunnane, S., Galli, C., Gehbremeskel, K., Linseisen, F., Lloyd-Smith, J., Parkington, J. (2000). Evidence for the unique function of docosahexaenoic acid (DHA) during the evolution of the modern hominid brain. Lipids, 34, S39-S47.

Crouse, C., Kelley, R., Trushenski, J., and Lydy, M. (2013). Use of alternative lipids and finishing feeds to improve nutritional value and food safety of hybrid striped bass. Aquaculture, 408–409, 58–69.

Dalla Valle, A., Iriti, M., Faoro, F., Berti, C., and Ciappellano, a. S. (2008). In vivo prion protein intestinal update in fish. APMIS, 116, 173–180.

Daly, P. (1981, November). Agricultural employment: has the decline ended? Monthly Labor Review, 11–17.

Darko, F., Quagrainie, K., and Chenyambuga, S. (2016). Consumer preferences for farmed tilapia in Tanzania: a choice experiment analysis. Journal of Applied Aquaculture, 28, 131–143.

Darwin, C. (1879). What Mr. Darwin Saw in His Voyage Round the World in the Ship "Beagle". New York: Harper and Brothers.

Davidson, K., Pan, M., Hu, W., and Poerwanto, D. (2012). Consumers' willingness to pay for aquaculture fish products vs. wild-caught seafood: a case study in Hawaii. Aquaculture Economics and Management, 16, 136–154.

Dawkins, R. (1995). River Out of Eden: A Darwinian View of Life. London: Weidenfeld and Nicolson.

Delwiche, J., Liggett, R., and Wallat, G. (2006). Consumer perception of cultured yellow perch (*Perca flavescens*) and its market competitors. Journal of Food Science, 71, S579-S582.

Denton, J., and Mellor, D. (n.d.). Cost and yield comparison of ready to cook chicken products. College Station, TX: Texas Agriculture Extension Service, Texas A & M University. Retrieved July 10, 2016, from http://posc.tamu.edu/wp-content/uploads/sites/20/2012/08/l-2290.pdf

Domingo, J., and Bocio, A. (2007). Levels of PCDD/PCDF and PCBs in edible marine species and human intake: a literature review. Environment International, 33, 397–405.

Done, H., and Halden, R. (2015). Reconnaissance of 47 antibiotics and associated microbial risks in seafood sold in the United States. Journal of Hazardous Materials, 282, 10–17.

Doreau, M., Corson, M., and Wiedermann, S. (2012). Water use by livestock: a global perspective for a regional issue? Animal Frontiers, 2, 9–16.

Dulsrud, A., Norberg, H., and Lenz, T. (2006). Too much or too little information? The importance of origin and traceability for consumer trust in seafood in Norway and Germany. In J. Luten, C. Jacobsen, K. Bekaert, A. Saebo, and J. Oehlenschlager, Seafood Research from Fish to Dish (pp. 213–228). Wageningen, the Netherlands: Wageningen Academic Publishers.

Eljarrat, E., Caixach, J., and Rivera, J. (2002). Determination of PCDDs and PCDFs in different animal feed ingredients. Chemosphere, 46, 1403–1407.

Engle, C., and Stone, N. (2013). Competitiveness of U.S. aquaculture within the current U.S. regulatory framework. Journal of Aquaculture Economics and Management, 17, 251–280.

Environmental Working Group. (2003). PCBs in farmed salmon: factor methods, unnatural results. Washington, D.C.: Environmental Working Group.

Equality Now. (2015). Ending Sex Discrimination in the Law. New York: Equality Now.

European Food Safety Authority. (2010). Results of the monitoring of dioxin levels in food and feed. EFSA Journal, 8, 1385.

FAO Regional Office for Asia and the Pacific. (2007). An overview of the impact of the tsunami on selected coastal fisheries resources in Sri Lanka and Indonesia. Bangkok, Thailand: Food and Agriculture Organization of the United Nations.

Federoff, N., Battisti, D., Beachy, R., Cooper, P., Fischhoff, D., Hodges, C., Knauf, V.C., Lobell, D., Mazur, B.J., Molden, D., Reynolds, M.P., Ronald, P.C., Rosegrant, M.W., Sanchez, P.A, Vonshak, A., and Zhu, J.-K. (2010). Radically rethinking agriculture for the 21st century. Science, 327, 833–834.

Fiedler, J., Lividini, K., Drummond, E., and Thilsted, S. (2016). Strengthening the contribution of aquaculture to food and nutrition security: the potential of a vitamin A-rich, small fish in Bangladesh. Aquaculture, 452, 291–303.

Fieldstadt, E. (2016, August 14). Reality check: is America already 'great' or in rapid decline? Retrieved July 1, 2017, from NBC News: http://www.nbcnews.com/news/us-news/reality-check-america-already-great-or-rapid-decline-n617101

Fish Health Section of the American Fisheries Society. (2016). Retrieved April 30, 2017, from Suggested procedures for the detection and identification of certain finfish and shellfish pathogens: http://www.afs-fhs.org/bluebook/bluebook-index.php

Fletcher, G., and Hew, C. (2002). Transgenic salmon for aquaculture. In Y. Kitagawa, T. Matsuda, and S. Iijima, Animal Cell Technology: Basic and Applied Aspects (pp. 101–105). Dordrecht: Springer Netherlands.

Food and Agriculture Organization. (1986). The production of fish meal and oil, FAO Fisheries Technical Paper No. 142. Rome: Food and Agriculture Organization of the United Nations. Retrieved June 23, 2016, from http://www.fao.org/3/a-x6899e/X6899E00.HTM

Food and Agriculture Organization. (1988). Definition of aquaculture, Seventh Session of the IPFC Working Party of Expects on Aquaculture, IPFC/WPA/WPZ, p. 1–3, RAPA/FAO, Bangkok.

Food and Agriculture Organization. (2003). World Agriculture: Towards 2015/2030. Rome, Italy: Food and Agriculture Organization of the United Nations.

Food and Agriculture Organization. (2008). State of World Fisheries and Aquaculture 2008. Rome: Food and Agriculture Organization of the United Nations.

Food and Agriculture Organization. (2011). Technical guidelines on aquaculture certification. Rome: Food and Agriculture Organization of the United Nations.

Food and Agriculture Organization. (2012). State of World Fisheries and Aquaculture. Rome: Food and Agriculture Organization of the United Nations.

Food and Agriculture Organization. (2014). The State of World Fisheries and Aquaculture. Rome: Food and Agriculture Organization of the United Nations.

Food and Agriculture Organization. (2016a, December 20). Food Supply – Livestock and Fish Primary Equivalent. Retrieved January 9, 2017, from FAOSTAT: http://www.fao.org/faostat/en/#data/CL

Food and Agriculture Organization. (2016b). The State of World Fisheries and Aquaculture. Rome: Food and Agriculture Organization of the United Nations.

Food and Agriculture Organization. (2016c). National Aquaculture Legislation Overview, Norway. Retrieved November 11, 2016, from http://www.fao.org/fishery/legalframework/nalo_norway/en

Food and Agriculture Organization. (2016d). National Aquaculture Legislation Overview, India. Retrieved November 11, 2016, from http://www.fao.org/fishery/legalframework/nalo_india/en

Food and Agriculture Organization. (2016e). National Aquaculture Legislation Overview, Viet Nam. Retrieved November 11, 2016, from http://www.fao.org/fishery/legalframework/nalo_vietnam/en

Food and Agriculture Organization. (2016f). National Aquaculture Legislation Overview, Bangladesh. Retrieved November 11, 2016, from http://www.fao.org/fishery/legalframework/nalo_bangladesh/en

Food and Agriculture Organization. (2016g). National Aquaculture Legislation Overview, Republic of Korea. Retrieved November 11, 2016, from http://www.fao.org/fishery/legalframework/nalo_korea/en#tcNB00F3

Food and Agriculture Organization. (2016h). National Aquaculture Legislation Overview, Egypt. Retrieved November 11, 2016, from http://www.fao.org/fishery/legalframework/nalo_egypt/en

Food and Agriculture Organization. (2016i). National Aquaculture Legislation Overview, The Philippines. Retrieved November 11, 2016, from http://www.fao.org/fishery/legalframework/nalo_philippines/en

Food and Agriculture Organization. (2016j, April 26). Meat and Meat Products. Retrieved July 4, 2017, from Agriculture and Consumer Protection Department, Animal Production and Health: www.fao.org/ag/againfo/themes/en/meat/home.html

Food and Agriculture Organization. (2016k). Scoping study on decent work and employment in fisheries and aquaculture: issues and actions for discussion and programming. Rome: Food and Agriculture Organization of the United Nations.

Food and Agriculture Organization. (2017a). Global Aquaculture Production 1950–2014. Retrieved January 9, 2017, from Fisheries Global Information System: http://www.fao.org/fishery/statistics/global-aquaculture-production/query/en

Food and Agriculture Organization. (2017b). Global Production Statistics 1950–2014. Retrieved January 9, 2017, from Fisheries Global Information System: http://www.fao.org/fishery/statistics/global-production/query/en

Food and Agriculture Organization. (2017c). Global Aquaculture Production 1950–2015. Rome, Italy. Retrieved May 7, 2017, from www.fao.org/figis/servlet/TabSelector

Food and Agriculture Organization. (2017d). National Aquaculture Sector Overview, Vietnam. Retrieved June 17, 2017, from http://www.fao.org/fishery/countrysector/naso_vietnam/en

Food and Agriculture Organization. (2017e). National Aquaculture Sector Overview, United States of America. Retrieved June 17, 2017, from http://www.fao.org/fishery/countrysector/naso_usa/en

Food and Agriculture Organization. (2017f). National Aquaculture Sector Overview, Philippines. Retrieved June 17, 2017, from http://www.fao.org/fishery/countrysector/naso_philippines/en

Food and Agriculture Organization. (2017g). National Aquaculture Sector Overview, Republic of Korea. Retrieved June 17, 2017, from http://www.fao.org/fishery/countrysector/naso_korea/en

Food and Agriculture Organization. (2017h). National Aquaculture Sector Overview, Norway. Retrieved June 17, 2017, from http://www.fao.org/fishery/countrysector/naso_norway/en

Food and Agriculture Organization. (2017i). National Aquaculture Sector Overview, Egypt. Retrieved June 17, 2017, from http://www.fao.org/fishery/countrysector/naso_egypt/en

Food and Agriculture Organization. (2017j). National Aquaculture Sector Overview, Chile. Retrieved June 17, 2017, from http://www.fao.org/fishery/countrysector/naso_chile/en

Food and Agriculture Organization. (2017k). National Aquaculture Legislation Overview, China. Retrieved June 5, 2017, from http://www.fao.org/fishery/legalframework/nalo_china/en

Food and Agriculture Organization. (2017l). Food Supply – Livestock and Fish Primary Equivalent. Retrieved January 27, 2018, from FAOSTAT: http://www.fao.org/faostat/en/#data/CL

Food and Agriculture Organization. (2017m, July 4). Food Balance Sheets. Retrieved from FAOSTAT: www.fao.org/faostat/en/#data/FBS

Food and Agriculture Organization. (2017n). State of Food Security and Nutrition in the World. Rome, Italy: Food and Agriculture Organization of the United Nations.

Food and Agriculture Organization. (2018a). National Aquaculture Sector Overview – China. Retrieved February 10, 2018, from http://www.fao.org/fishery/countrysector/naso_china/en

Food and Agriculture Organization. (2018b). National Aquaculture Sector Overview – Indonesia. Retrieved February 10, 2018, from http://www.fao.org/fishery/countrysector/naso_indonesia/en

Food and Agriculture Organization. (2018c). National Aquaculture Sector Overview – India. Retrieved February 10, 2018, from http://www.fao.org/fishery/countrysector/naso_india/en

Food and Agriculture Organization. (2018d). National Aquaculture Sector Overview – Bangladesh. Retrieved February 10, 2018, from http://www.fao.org/fishery/countrysector/naso_bangladesh/en

Food and Agriculture Organization. (2018e). National Aquaculture Legislation Overview – Indonesia. Retrieved August 6, 2018, from http://www.fao.org/fishery/legalframework/nalo_indonesia/en

Food and Agriculture Organization. (2018f). National Aquaculture Legislative Overview, Chile. Retrieved August 6, 2018, from http://www.fao.org/fishery/legalframework/nalo_chile/en

Food and Agriculture Organization. (2018g). Global Aquaculture Production. Fishery Statistical Collections. Rome, Italy.

Food and Agriculture Organization/International Fund for Agricultural Development/World Food Programme. (2015). The State of Food Insecurity in the World. Rome: Food and Agriculture Organization of the United Nations.

Food and Agriculture Organization/World Health Organization. (2003). Position Paper on Dioxins and Dioxin-like PCBs. Rome: FAO/WHO Joint Office.

Food and Agriculture Organization/World Health Organization. (2011). Report of the joint FAO/WHO expert consultation on the risks and benefits of seafood consumption. Rome: Food and Agriculture Organization of the United Nations.

Forabosco, F., Lohmus, M., Rydhmer, L., and Sundstrom, L. (2013). Genetically modified farm animals and fish in agriculture: a review. Livestock Science, 153, 1–9.

Forbes, K., and Broadhead, J. (2007). The role of coastal forests in the mitigation of tsunami impacts. Bangkok: Food and Agriculture Organization of the United Nations Regional Office for Asia and the Pacific.

Ford, J., and Myers, R. (2008). A global assessment of salmon aquaculture impacts on wild salmonids. PLoS Biology, 6, 411–417.

Froehlich, H., Gentry, R., Rust, M., Grimm, D., and Halpern, B. (2017). Public perceptions of aquaculture: evaluating spatiotemporal patterns of sentiment around the world. PLOS One. doi:10.1371/journal.pone.0169281

Froese, R., and Pauly, D. (2017, February). FishBase. Retrieved February 12, 2017, from FishBase

Gallardi, D. (2014). Effects of bivalve aquaculture on the environment and their possible mitigation: a review. Fisheries and Aquaculture Journal, 5, 1000105.

Gerber, P., Steinfeld, H., Benderson, B., Mottet, A., Opio, C., Dijkman, J., Falcucci, A., and Tempio, G. (2013). Tackling climate change through livestock – a global assessment of emissions and mitigation opportunities. Rome: Food and Agriculture Organization of the United Nations.

Gladwell, M. (1995, May 19). DNA testing itself no longer on trial; Simpson defense team tactics reflect procedure's acceptance. The Washington Post, p. A01.

Gladwell, M. (2000). The Tipping Point: How Little Things Can Make a Big Difference. New York: Little, Brown, and Company.

Glaropoulos, A., Papadakis, V., Papadakis, I., and Kentouri, M. (2012). Escape-related behavior and coping ability of sea bream due to food supply. Aquaculture International, 20, 965–979.

Gorton's of Gloucester. (n.d.). Gorton's of Gloucester. Retrieved July 1, 2017, from We are Gorton's: https://www.gortons.com/we-are-gortons/

GovTrack. (n.d.-a). S. 1195 (109th): National Offshore Aquaculture Act of 2005. Retrieved from GovTrack: https://www.govtrack.us/congress/bills/109/s1195

GovTrack. (n.d.-b). H.R. 2010 (110th): National Offshore Aquaculture Act of 2007. Retrieved July 1, 2017, from GovTrack: https://www.govtrack.us/congress/bills/110/hr2010

GovTrack. (n.d.-c). S. 1609 (110th): National Offshore Aquaculture Act of 2007. Retrieved from GovTrack: https://www.govtrack.us/congress/bills/110/s1609

GovTrack. (n.d.-d). H.R. 4364 (111th): National Sustainable Offshore Aquaculture Act of 2009. Retrieved from GovTrack: https://www.govtrack.us/congress/bills/111/hr4363

GovTrack. (n.d.-e). GovTrack. Retrieved July 1, 2017, from H.R. 2373 (112th): National Sustainable Offshore Aquaculture Act of 2011.: https://www.govtrack.us/congress/bills/112/hr2373

Green, K. (2004). The great salmon panic of 2004. Fraser Forum, March, 20–22.

Greenberg, P. (2010). Four Fish: The Future of the Last Wild Food. New York: Penguin Group.

Gresham, C., and Taylor, L. (2015, December 29). Seafood toxicity. Retrieved August 27, 2017, from Medscape: emedicine.medscape.com/article/1011549-overview

Gross, M. (1998). One species with two biologies: Atlantic salmon (*Salmo salar*) in the wild and in aquaculture. Canadian Journal of Fisheries and Aquatic Sciences, 131–144.

Gupta, S. (2014, May 11). Salmon in the Sea. [Television Program]. 60 Minutes. CBS Interactive, Inc. Retrieved from https://www.cbsnews.com/news/saving-wild-salmon/

Halverson, A. (2010). An Entirely Synthetic Fish: How Rainbow Trout Beguiled America and Overran the World. New Haven, CT: Yale University Press.

Halwart, M., and Gupta, M. (2004). Culture of Fish in Rice Fields. Rome, Italy: Food and Agriculture Organization of the United Nations, The WorldFish Center.

Hamilton, S. (2013). Assessing the role of commercial aquaculture in displacing mangrove forest. Bulletin of Marine Science, 89, 585–601.

Hardin, G. (1968). The Tragedy of the Commons. Science, 162, 1243–1248.

Hardin, G. (1998). Extensions of "The Tragedy of the Commons". Science, 280, 682–683.

Hardy, N. (2009). Innocence Project. In H. Greene, and S. Gabbidon, Encyclopedia of Race and Crime. Sage Publications.

Harrell, R., Kerby, J., and Minton, R. (1990). Culture and Propagation of Striped Bass and its Hybrids. Bethesda, MD: American Fisheries Society.

Hawke, J. (2015). Retrieved November 5, 2016, from https://srac.tamu.edu/serveFactSheet/124

Health Canada. (2016a, May 19). AquAdvantage Salmon. Retrieved January 22, 2017, from http://www.hc-sc.gc.ca/fn-an/gmf-agm/appro/aquadvantage-salmon-saumon-eng.php

Health Canada. (2016b, May 19). Frequently Asked Questions: AquAdvantage Salmon. Retrieved January 22, 2017, from http://www.hc-sc.gc.ca/fn-an/gmf-agm/appro/aquadvantage-salmon-saumon-faq-eng.php

Henderson, R. (2015, December). Industry employment and output projections to 2024. Retrieved June 30, 2017, from Monthly Labor Review: https://www.bls.gov/opub/mlr/2015/article/industry-employmet-and-output-projections-to-2024.htm

Heshmati, A., Karami-Momtaz, J., Nili-Ahmadabadi, A., and Ghadimi, S. (2017). Dietary exposure to toxic and essential trace elements by consumption of wild and farmed carp (*Cyprinus carpio*) and Caspian kutum (*Rutilus frisii kutum*) in Iran. Chemosphere, 173, 207–215.

Hicks, D., Pivarnik, L., and McDermott, R. (2008). Consumer perceptions about seafood: an internet survey. Journal of Foodservice, 19, 213–226.

High Liner Foods. (2014). High Liner Foods. Retrieved July 1, 2017, from Retail: www.highlinerfoods.com/en/home/our-brands/retail/default.aspx

Hilbeck, A., Binimelis, R., Defarge, N., Steinbrecher, R., Szekacs, A., Wickson, F., Antoniou, M., Bereano, P.L., Clark, E.A., Hansen, M., Novotny, E., Heinemann, J., Meyer, H., Shiva, V., and Wynne, B. (2015). No scientific consensus on GMO safety. Environmental Sciences Europe, 27, 4.

Hishamunda, N., Ridler, N., and Martone, E. (2014). Policy and governance in aquaculture. Rome: Food and Agriculture Organization of the United Nations.

Hites, R., Foran, J., Carpenter, D., Hamilton, M., Knuth, B., and Schwager, S. (2004). Global assessment of organic contaminants in farmed salmon. Science, 303, 226–229.

Hoekstra, A., and Chapagain, A. (2007). Water footprints of nations: water use by people as a function of their consumption pattern. Water Resource Management, 21, 35–48.

Hoekstra, A., and Mekonnen, M. (2012). The water footprint of humanity. PNAS, 109, 3232–3237.

Hoffman, M. (2000). Empathy and Moral Development. Cambridge, UK: Cambridge University Press.

Holland, D., and Wessells, C. (1998). Predicting consumer preferences for fresh salmon: the influence of safety inspection and production method attributes. Agricultural and Resource Economics Review, 1–14.

Huber, S., Hansen, L., Rasmussen, M., and Kass, H. (2016). Characterization of Danish waters with EO and modelling for aquaculture site selection. Living Planet Symposium (p. 262). Prague, Czech Republic: European Space Agency.

Hulata, G. (2011). Genetic manipulations in aquaculture: a review of stock improvement by classical and modern technologies. Genetica, 111, 155–173.

Huntingord, F. (2008). Animal welfare in aquaculture. In K. Culver, and D. Castle, Aquaculture, Innovation and Social Transformation (pp. 21–33). Dordrecht, The Netherlands: Springer.

Huss, H. (1994). Assurance of seafood quality, FAO Fisheries Technical Paper 334. Rome: Food and Agriculture Organization of the United Nations.

Huwe, J., and Archer, J. (2013). Dioxin congener patterns in commercial catfish from the United States and the indication of mineral clays as the potential source. Food Additives and Contaminants Part A, 30, 331–338.

Inman, and Beck. (2011). DNA evidence. In W. Chambliss, Key Issues in Crime and Punishment: Courts, Law, and Justice. Thousand Oaks, CA: SAGE Publications.

International Monetary Fund. (2017, April). Database--WEO Groups and Aggregates Information. Washington, D.C., USA. Retrieved May 29, 2017, from https://www.imf.org/external/pubs/ft/weo/2017/01/weodata/groups.htm#lac

Iwamoto, M., Ayers, T., Mahon, B., and Swerdlow, D. (2010). Epidemiology of seafood-associated infections in the United States. Clinical Microbiology Reviews, 23, 399–411.

Jackson, A. (2013). How much fish is consumed in aquaculture? Global Aquaculture Advocate, January/February, 28–31. Retrieved

July 10, 2016, from http://www.gaalliance.org/mag/2013/Jan-Feb/download.pdf

Jacobs, M., Covaci, A., and Schepens, P. (2002). Investigation of selected persistent organic pollutants in farmed Atlantic salmon (*Salmo salar*), salmon aquaculture feed, and fish oil components of the feed. Environmental Science and Technology, 36, 2797–2805.

Jennings, S., Stentiford, G., Leocadio, A., Jeffery, K., Metcalfe, J., Katsiadaki, I., Auchterlonie, N.A., Mangi, S.C., Pinnegar, J.K., Ellis, T., Peeler, E.J., Luisetti, T., Baker-Austin, C., Brown, M., Catchpole, T.L., Clyne, F.J., Dye, S.R., Edmonds, N.J., Hyder, K., Lee, J., Lees, D.N., Morgan, O.C., O'Brien, C.M., Oidtmann, B., Posen, P.E., Ribeiro Santos, A., Taylor, N.G.H., Turner, A.D., Townhill, B.L., Verner-Jeffreys, D.W. (2016). Aquatic food security: insights into challenges and solutions from an analysis of interactions between fisheries, aquaculture, food safety, human health, fish and human welafre, economy and environment. Fish and Fisheries, 17, 893–938.

Jensen, I., Dempster, T., Thorstad, E., Uglem, I., and Fredheim, A. (2010). Escapes of fishes from Norwegian sea-cage aquaculture: causes, consequences and prevention. Aquaculture Environment Interactions, 1, 71–83.

Johnson, B., and Bosworth, B. (2012). Investigational New Animal Drug (INAD) exemptions and the National INAD Program (NIP), SRAC Publication No. 4709. Retrieved July 30, 2018, from https://agrilifecdn.tamu.edu/fisheries/files/2013/09/SRAC-Publication-No.-4709-Investigational-New-Animal-Drug-INAD-Exemptions-and-the-National-INAD-Program-NIP.pdf

Jones, S., Bruno, D., Madsen, L., and Peeler, E. (2015). Disease management mitigates risk of pathogen transmission from maricultured salmonids. Aquaculture Environment Interactions, 6, 119–134.

Joram, A., and Kapute, F. (2016). Sensory evaluation of wild-captured and pond-raised tilapias in Malawi. African Journal of Food Science, 10, 238–242.

Joyce, A., and Satterfield, T. (2010). Shellfish aquaculture and First Nations sovereignty: the quest for sustainable development in contested sea space. Natural Resources Forum, 34, 106–123.

Kac, E. (2000). GFP Bunny. Retrieved January 16, 2017, from http://www.ekac.org/gfpbunny.html#gfpbunnyanchor

Kapetsky, J., Hill, J. W., and Evans, D. (1990). Assessing potential for aquaculture development with a geographic information system. Journal of the World Aquaculture Society, 21, 241–249.

Karimi, R., Fitzgerald, T., and Fisher, N. (2016). A quantitative synthesis of mercury in commercial seafood and implications for exposure in the United States. Environmental Health Perspectives, 120, 1512–1519.

Karlsen, K., and Senneset, G. (2006). Traceability: simulated recall of fish products. In J. Luten, C. Jacobsen, K. Bekaert, A. Saebo, and J. Oehlenschlager, Seafood research from fish to dish (pp. 251–262). Wageningen, the Netherlands: Wageningen Academic Publishers.

Karlsen, K., Andreassen, O., and Hersoug, B. (2015). From controvsery to dialog in aquaculture. Tromso: Nofima.

Keane, E. (2016, November 19). From truthiness to post-truth, just in time for Donald Trump: Oxford Dictionaries' word of the year should scare the hell out of you. Retrieved from Salon: https://www. salon.com/2016/11/19/from-truthiness-to-post-truth-just-in-time-for-donald-trump-oxford-dictionaries-word-of-the-year-should-scare-the-hell-out-of-you/

Kecinski, M., Messer, K., Knapp, L., and Shirazi, Y. (2017). Consumer preferences for oyster attributes: field experiments on brand, locality, and growing method. Agricultural and Resource Economics Review, 1–23.

Kibenge, M., Iwamoto, T., Wang, Y., Morton, A., Routledge, R., and Kibenge, F. (2016). Discovery of variant infectious salmon anaemia virus (ISAV) of European genotype in British Columbia, Canada. Virology Journal, 13, 3–19.

Knapp, G. (2007). Implications of aquaculture for wild fisheries: the case of Alaska wild salmon. In R. Arthur, and J. Neirentz, Global Trade Conference on Aquaculture (pp. 239–245). Rome, Italy: Food and Agriculture Organization of the United Nations.

Knapp, G., and Rubino, M. (2016). The political economics of marine aquaculture in the United States. Reviews in Fisheries Science and Aquaculture, 24, 213–229.

Kobayashi, M., Msangi, S., Batka, M., Vannuccini, S., Dey, M., and Anderson, J. (2015). Fish to 2030: the role and opportunity for aquaculture. Aquaculture Economics and Management, 19, 282–300.

Kolar, C., Chapman, D., Courtenay, W., Housel, C., Williams, J., and Jennings, D. (2005). Asian carps of the genus Hypophthalmichthys (Pisces, Cyprinidae) – a biological synopsis and environmental risk assessment. Washington, DC: U.S. Fish and Wildlife Service.

Krause, V. (2011a, March 14). David Suzuki's fish story. Retrieved February 12, 2017, from Financial Post: http://business.financialpost.com/fp-comment/suzukis-fish-story

Krause, V. (2011b, January 26). Packard's push against B.C. salmon. Retrieved July 1, 2017, from The Financial Post: business.financial post.com/opinion/packards-push-against-b-c-salmon/wcm/ad8c6f63–56e5–8de4-e3532e1b023a

Krkošek, M. (2017). Population biology of infectious diseases shared by wild and farmed fish. Canadian Journal of Fisheries and Aquatic Science, 74, 620–628.

Kurlansky, M. (1997). Cod: A Biography of the Fish that Changed the World. New York: Penguin Group.

Lafferty, K. (1999). The evolution of trophic transmission. Parasitology Today, 15, 111–115.

Lafferty, K., and Hofmann, E. (2016). Marine disease impacts, diagnosis, forecasting, management and policy. Philosophical Transactions of the Royal Society B. doi:10.1098/rstb.2015.0200

Lafferty, K., Harvell, C., Conrad, J., Friedman, C., Kent, M., Kuris, A., Powell, E.N., Rondeau, D., and Saksida, S. (2015). Infectious diseases affect marine fisheries and aquaculture economics. Annual Reviews in Marine Science, 7, 471–496.

Laland, K., Brown, C., and Krause, J. (2003). Learning in fishes: from three-second memory to culture. Fish and Fisheries, 199–202.

LaPatra, S., and Foott, J. (2006). Disease transmission from aquaculture to wild stocks: a case study in risk evaluation. World Aquaculture (September), 53–55.

Lee, O., Green, J., and Tyler, C. (2015). Transgenic fish systems and their application in ecotoxicology. Critical Reviews in Toxicology, 45, 124–141.

Lessler, J., and Ranells, N. (2007). Grower guidelines for poultry and fowl processing. Pittsboro, NC: North Carolina Cooperative Extension, North Carolina State University. Retrieved July 10, 2016, from http://chatham.ces.ncsu.edu/growingsmallfarms/GrowerGuidelines.pdf

Lien, S., Koop, B., Sandve, S., Miller, J., Kent, M., Nome, T., Hvidsten, T.R., Leong, J.S> Minkley, D.R., Zimin, A., Grammes, F., Grove, H., Gjuvsland, A., Walenz, B., Hermansen, R.A., von Shalburg, K., Rondeau, E.B., De Genova, A., Samy, J.K., Olav Vik, J., Viegland, M.D., Caler, L., Grimholt, U., Jentoft, S., Våge, D.I., de Jong, P., Moen, T., Baranski, M. Palti, Y., Smith, D.R., Yorke, J.A., Nederbragt, A.J., Tooming-Klunderud, A., Jakobsen, K.S., Jiang, X., Fan, D., Hu, Y., Liberles, D.A., Vidal, R., Iturra, P., Jones, S.J., Jonassen, I., Maass, A. Omholt, S.W., and Davidson, W.S. (2016). The Atlantic salmon genome provides insights into rediploidization. Nature, 533, 200–205.

Lloyd, W. (1833). Two lectures on the checks to population. Oxford, UK: Oxford University.

Loredana, I., Novoa, B., Dalla, A., Cardone, F., Aranguren, R., Sbriccoli, M., Bevivino, S., Iriti, M., Liu, Q., Vetrugno, V., Lu, M., Faoro, F., Ciappellano, S., Figueras, A., and Pocchiari, M. (2006). Scrapie infectivity is quickly cleared in tissues of orally-infected farmed fish. BMC Veterinary Research, 2, 21–27.

Lovatelli, A. (2008). Capture-based aquaculture: global overview. Rome: Food and Agriculture Organization of the United Nations. Retrieved June 22, 2016, from http://www.fao.org/docrep/011/i0254e/i0254e00.htm

Love, C., and Stamps, C. (2008). Animals: a visual encyclopedia. London: Dorling Kindersley Publishing.

Love, D., Rodman, S., Neff, R., and Nachman, K. (2011). Veterinary drug residues in seafood inspected by the European Union, United States, Canada, and Japan from 2000 to 2009. Environmental Science and Technology, 45, 7232–7240.

Lovell, S., and Stone, S. (2005). The economic impacts of aquatic invasive species: a review of the literature. Washington, DC: U.S. Environmental Protection Agency.

Magee, J.G. (1942). High Flight. Poems of Faith and Freedom exhibition. Library of Congress, Washington, D.C.

Male, R., Lorens, J., Nerland, A., and Slinde, E. (1993). Biotechnology in aquaculture, with special reference to transgenic salmon. Biotechnology and Genetic Engineering Reviews, 11, 33–56.

Malisch, R., and Kotz, A. (2014). Dioxins and PCBs in feed and food – review from European perspective. Science of the Total Environment, 491–492, 2–10.

Mangrove Action Project. (n.d.). Shrimp Farming. Retrieved February 24, 2018, from http://mangroveactionproject.org/shrimp-farming/

Mansfield, B. (2011). "Modern" industrial fisheries and the crisis of over-fishing. In R. Peet, P. Robbins, and M. Watts, Global Political Ecology (pp. 84–99). London: Routledge.

Mapes, L. (2017, November 14). Escaped Atlantic salmon have disappeared form Puget Sound, but legal Fight begins. Seattle Times. Seattle.

Mapes, L., and Bernton, H. (2017, August 22). Please go fishing, Washington state says after farmed Atlantic salmon escape broken net. Seattle Times. Retrieved September 10, 2017, from www.seattletimes.com/seattle-news/environment/oops-after-accidental-release-of-atlantic-salmon-fisherman-being-told-catch-as-many-as-you-want/

Marine Aquaculture Task Force. (2007). Sustainable marine aquaculture: fulfilling the promise, managing the risks. Takoma Park, Maryland: Marine Aquaculture Task Force.

Marques, A., Lourenco, H., Nunes, M., Roseiro, C., Santos, C., Barranco, A., Rainieri, S., Langerholc, T., Cencic, A. (2011). New tools to assess toxicity, bioavailability and uptake of chemical contaminants in meat and seafood. Food Research International, 510–522.

Martin, D. (1998, January 21). To boldly go where fish haven't; testing feasibility of piscine farms for space voyages. Retrieved July 8, 2017, from New York Times Online: www.nytimes.com/1998/01/21/nyregion/boldly god-where-fish-haven-t-feasibility-piscine-farms-for-space-voyages.html

Martins, C., Eding, E., Verdegem, M., Heinsbroek, L., Schneider, O., Blancheton, J., Roque d'Orbcastel, E., and Verreth, J. (2010). New developments in recirculating aquaculture systems in Europe: a perspective on environmental sustainability. Aquacultural Engineering, 43, 83–93.

Marzec, R. (2004). The Greenwood Encyclopedia of American Regional Cultures: the Mid-Atlantic Region. Santa Barbara, CA: ABC-CLIO.

Maule, A., Gannam, A., and Davis, J. (2007). Chemical contaminants in fish feeds used in federal salmonid hatcheries in the USA. Chemosphere, 67, 1308–1315.

McClenachan, L. N., Al-Abdulrazzak, D., Witkin, T., Fisher, K., and Kittinger, J. (2014). Do community supported fisheries (CSFs) improve sustainability? Fisheries Research, 157, 62–69.

McDonough, S., Gallardo, W., Berg, H., Trai, N., and Yen, N. (2014). Wetland ecosystem service values and shrimp aquaculture relationships in Can Gio, Vietnam. Ecological Indicators, 46, 201–213.

McKindsey, C., Anderson, M., Barnes, P., Courtenay, S., Landry, T., and Skinner, M. (2006). Effects of shellfish aquaculture on fish habitat. Ottawa, Canada: Canadian Science Advisory Secretariat.

Miller, M., and Vincent, E. (2008). Rapid natural selection for resistance to an introduced parasite of rainbow trout. Evolutionary Applications, 1, 336–341.

Milner-Gulland, E., Bennett, E., and Group, S. 2. (2003). Wild meat: the bigger picture. Trends in Ecology and Evolution, 18, 351–357.

Minamata Disease Municipal Museum. (2007, December). Minamata disease: its history and lessons. Retrieved March 14, 2017, from http://www.minamata195651.jp/pdf/kyoukun_en/kyoukun_eng_all.pdf

Mitchell, A., and Kelly, A. (2006). The public sector role in the establishment of grass carp in the United States. Fisheries, 31, 113–121.

Molden, D., Oweis, T., Steduto, P., Bindraban, P., Hanjra, M., and Kijne, J. (2010). Improving agricultural water productivity: between optimism and caution. Agricultural Water Management, 97, 528–535.

Monterrey Bay Aquarium. (2015, October 30). 2016 Seafood Watch Standard for Aquaculture. Retrieved June 22, 2016, from Seafood Watch: https://www.seafoodwatch.org/-/m/sfw/pdf/standard%20 revision%20reference/2015%20standard%20revision/mba_seafood watch_aquaculture%20criteria_final.pdf?la=en

Moretti, V., Turchini, G., Bellagamba, F., and Caprino, F. (2003). Traceability issues in fishery and aquaculture products. Veterinary Research Communications, 27, Supp 1, 497–505.

Mozaffarian, D., and Rimm, E. (2006). Fish intake, contaminants, and human health: evaluating the risks and the benefits. Journal of the American Medical Association, 296, 1885–1899.

Muir, J. (2015). Fuel and energy use in the fisheries sector. Rome, Italy: Food and Agriculture Organization of the United Nations.

Murray, G., Wolff, K., and Patterson, M. (2017). Why eat fish? Factors influencing seafood consumer choices in British Columbia, Canada. Ocean and Coastal Management, 144, 16–22.

National Aeronautics and Space Administration. (2012, July 25). A Fish Friendly Facility for the International Space Station. Retrieved July 8, 2017, from International Space Station: https://www.nasa.gov/mis sion_pages/station/research/news/aquatic.html

National Research Council. (2003). Dioxins and dioxin-like compounds in the food supply: strategies to decrease exposure. Washington, D.C.: National Academies Press.

National Research Council. (2011). Nutrient Requirements of Fish and Shrimp. Washington, DC: National Academies Press.

Naylor, R., Goldburg, R., Primavera, J., Kautskhy, N., Beveridge, M., Clay, J., Folke, C., Lubchenco, J., Mooney, H., and Troell, M. (2000). Effect of aquaculture on world fish supplies. Nature, 405, 1017–1024.

Naylor, R., Hardy, R., Bureau, D., Chiu, A., Elliott, M., Farrell, A., Forster, I., Gatlin, D.M., Goldburg, R.J., Hua, K., and Nichols, P. (2009). Feeding aquaculture in an era of finite resources. PNAS, 106, 15103–15110.

Neori, A., and Nobre, A. (2012). Relationship between trophic level and economics in aquaculture. Aquaculture Economics and Management, 16, 40–67.

Nico, L., Maynard, E., Schofield, P., Cannister, M., Larson, J., Fusaro, A., and Neilson, M. (2014, June 26). Cyprinus carpio. Retrieved from USGS Nonindigenous Aquatic Species Database: http://nas.er.usgs.gov/queries/FactSheets.aspx?SpeciesID=4

Nijdam, D., Rood, T., and Westhoek, H. (2012). The price of protein: review of land use and carbond footprints from life cycle assessments of animal food products and their substitutes. Food Policy, 37, 760–770.

Nuflor – BRD and Foot Rot. (n.d.). Retrieved November 5, 2016, from Nuflor (florfenicol): http://www.nuflor.com/diseases/fr-frp.asp

Osmundsen, T., and Olsen, M. (2017). The imperishable controversy over aquaculture. Marine Policy, 76, 136–147.

Osmundsen, T., Almklov, P., and Tveteras, R. (2017). Fish farmers and regulators coping with the wickedness of aquaculture. Aquaculture Economics and Management, 21, 163–183.

Oxford University Press. (2016, June 18). Aquaculture. Retrieved from Oxford English Dictionary: www.oed.com

Padiyar, P., Subachri, W., Pamudi, Raharjo, S., Phillips, M., and Subasinghe, R. (2006). Recovery and sustainable development of aquaculture industry in tsunami affected Aceh and Nias provinces in Indonesia. FAO Aquaculture Newsletter #36. Rome, Italy. Retrieved from http://www.fao.org/tempref/docrep/fao/009/a0974e/A0974E08.pdf

Pandian, T., and Sheela, S. (1995). Hormonal induction of sex reversal in fish. Aquaculture, 138, 1–22.

Parker, R., and Tyedmers, P. (2014). Fuel consumption of global fishing fleets: current understanding and knowledge gaps. Fish and Fisheries, 16, 684–696.

Patterson, T. (2016, September 7). Research: media coverage of the 2016 election. Retrieved July 3, 2017, from Harvard Kennedy School Shorenstein Center on Media, Politics and Public Policy: https://shorensteincenter.org/research-media-coverage-2016-election/

Paul, D., Christensen, V., Dalsgaard, J., Froese, R., and Torres, F. (1998). Fishing down marine food webs. Science, 279, 860–863.

Pauly, D., Tyedmers, P., Froese, R., and Liu, L. (2001). Fishing down and farming up the food web. Conservation Biology in Practice, 2, 1–25.

Pauly, D., Watson, R., and Alder, J. (2005). Global trends in world fisheries: impacts on marine ecosystems and food security. Philosophical Transactions of the Royal Society B, 360, 5–12.

Pew Charitable Trusts. (2004, January 9). Global Assessment of Organic Contaminants in Farmed Salmon. Retrieved February 12, 2017, from http://www.pewtrusts.org/en/research-and-analysis/reports/2004/01/09/global-assessment-of-organic-contaminants-in-farmed-salmon

Pieniak, Z., Verbeke, W., Brunso, K., and Olsen, S. (2006). Consumer knowledge and interest in information about fish. In J. Luten, C. Jacobsen, K. Bekaert, A. Saebo, and J. Oehlenschlager, Seafood research from fish to dish (pp. 229–240). Wageningen, the Netherlands: Wageningen Academic Publishers.

Pieniak, Z., Vanhonacker, F., and Verbeke, W. (2013). Consumer knowledge and use of information about fish and aquaculture. Food Policy, 40, 25–30.

Pimental, D., McNair, S., Janecka, S., Wightman, J., Simmonds, C., O'Connell, C., Wong, E., Russel, L, Zern, J., Aquino, T, and Tsomondo, T. (2001). Economic and environmental threats of alien plant, animals, and microbe invasions. Agriculture, Ecosystems and Environment, 84, 1–20.

Piper, R., McElwain, I., Orme, L., McCraren, J., Fowler, L., and Leonard, J. (1986). Fish Hatchery Management. Washington, D.C.: U.S. Department of the Interior.

Population Fund of the United Nations. (2014). State of World Population. Rome: Population Fund of the United Nations.

Porteous, A. (2008). Dictionary of environmental science and technology. Hoboken, NJ: Wiley.

Prescott, S., Pawankar, R., Allen, K., Campbell, D., Sinn, J., Fiocchi, A., Ebisawa, M., Sampson, H.A., Beyer, K., Lee, B.-W. (2013). A global survey of changing patterns of food allergy burden in children. World Allergy Organization Journal, 6, 21–32.

Rabanal, H. (1988). History of Aquaculture. Rome: Food and Agriculture Organization of the United Nations.

Rentfrow, G. (2010). How much meat to expect from a carcass, a consumer's guide to purchasing freezer meats. Lexington, KY: Cooperative Extension Service, University of Kentucky. Retrieved July 10, 2016, from http://www2.ca.uky.edu/agcomm/pubs/asc/asc179/asc179.pdf

Riediger, N., Othman, R., Suh, M., and Moghadasian, M. (2009). A systemic review of the roles of n-3 fatty acids in health and disease. Journal of the American Dietetic Association, 109, 668–679.

Rombenso, A., Trushenski, J., Jirsa, D., and Drawbridge, M. (2016). Docosahexaenoic acid (DHA) and arachidonic acid (ARA) are essential to meet LC-PUFA requirements of juvenile California Yellowtail (*Seriola dorsalis*). Aquaculture, 463, 123–134.

Rosenberg, M. (2013, September 28). We're eating what? 9 Contaminants in US meat. Retrieved September 23, 2016, from OpEdNews.com: http://www.opednews.com/articles/We-re-Eating-What-9-Conta-by-Martha-Rosenberg-Animals_Animals_Beef_Cancer-130928-645.html

Rubino, M. (2008). Offshore aquaculture in the United States: economic considerations, implications, and opportunities. Silver Spring, MD: U.S. Department of Commerce.

Rumsey, G. (1993). Fish meal and alternate sources of protein in fish feeds, update 1993. Fisheries, 18, 14–19.

Ryder, J., Karunasagar, I., and Ababouch, L. (2014). Assessment and management of seafood safety and quality: current practices and emerging issues, FAO Fisheries and Aquaculture Technical Paper 574. Rome: Food and Agriculture Organization of the United Nations.

Salze, G., and Davis, D. (2015). Taurine: a critical nutrient for future fish feeds. Aquaculture, 437, 215–229.

Schlag, A. (2010). Aquaculture: an emerging issue for public concern. Journal of Risk Research, 13, 829–844.

SENCER. (2017, March 15). Do the benefits of aquaculture outweigh its negative impacts? Retrieved January 27, 2018, from KQED Education: https://ww2.kqed.org/education/2017/03/15/do-the-benefits-of-aquaculture-outweigh-its-negative-impacts/

Shackleton, P., and Hanlan, J. (2004). United Hatters, Cap, and Millinery Workers International Union. In R. Weir (Ed.), Historical Encyclopedia of American Labor. Santa Barbara, CA: ABC-CLIO.

Sheehan, M., Burke, T., Navas-Acien, A., Breysse, P., McGready, J., and Fox, M. (2014). Global methylmercury exposure from seafood consumption and risk of developmental neurotoxicity: a systematic review. Bulletin of the World Health Organization, 92, 254–269.

Sicherer, S., Munoz-Furlong, A., and Sampson, H. (2004). Prevalence of seafood allergy in the United States determined by a random telephone survey. Journal of Allergy and Clinical Immunology, 114, 159–165.

Simon Fraser University. (2010, March 17). Six to receive honorary SFU degrees. Retrieved May 1, 2017, from http://www.sfu.ca/archive-university-communications/media_releases/media_releases_archives/six-to-receive-honorary-sfu-degrees.html

Soto, D. (2009). Integrated mariculture, a global review. Rome, Italy: Food and Agriculture Organization of the United Nations.

Sparks, J., Schelly, R., Smith, W., Davis, M., Tchernov, D., Pieribone, V., and Gruber, D. (2014). The covert world of fish biofluorescence: a phylogenetically widespread and phenotypically variable phenomenon. PLOS One, http://dx.doi.org/10.1371/journal.pone.0083259

Staton, M., Edwards, H., Brisbin, I., Joanen, T., and McNease, L. (1990). Essential fatty acid nutrition of the American alligator (Alligator mississippiensis). Journal of Nutrition, 120, 674–685.

Stewart, C. (2006). Go with the glow: fluorescent proteins to light transgenic organisms. Trends in Biotechnology, 24, 155–162.

Stillwell, W., and Wassall, S. (2003). Docosahexaenoic acid: membrane properties of a unique fatty acid. Chemistry and Physics of Lipids, 126, 1–27.

Strochlic, N. (2015, February 16). The world's craziest anti-woman laws. Retrieved January 29, 2017, from The Daily Beast: http://www.the dailybeast.com/articles/2015/02/16/the-world-s-craziest-anti-women-laws.html

Stroud, R. (1986). Fish Culture in Fisheries Management. Bethesda, Maryland: American Fisheries Society.

Subasignhe, R., Soto, D., and Jia, J. (2009). Global aquaculture and its role in sustainable development. Reviews in Aquaculture, 1, 2–9.

Subasinghe, R., Arthur, J., Bartley, D., Silva, S. D., Halwart, M., Hishamunda, N., Mohan, C.V., and Sorgeloos, P. (2012). Farming the water for people and food. Rome/Bangkok: Food and Agriculture Organization of the United Nations/Network of Aquaculture Centres in Asia-Pacific.

Sveinsdottir, K., Martinsdottir, E., Green-Petersen, D., Hyldig, G., Schelvis, R., and Delahunty, C. (2009). Sensory characteristics of different cod products related to consumer preferences and attitudes. Food Quality and Preference, 20, 120–132.

Tacon, A., and Metian, M. (2008). Global overview on the use of fish meal and fish oil in industrially compounded aquafeeds: trends and future prospects. Aquaculture, 285, 146–158.

Tacon, A., and Metian, M. (2009). Fishing for feed or fishing for food: increasing global competition for small pelagic forage fish. AMBIO, 38, 294–302.

Tacon, A., and Metian, M. (2013). Fish matters: importance of aquatic foods in human nutrition and global food supply. Reviews in Fisheries Science, 21, 22–38.

Tacon, A., Metian, M., Turchini, G., and Silva, S. D. (2010). Responsible aquaculture and trophic level implications to global fish supply. Reviews in Fisheries Science, 18, 94–105.

Tacon, A., Hasan, M., and Metian, M. (2011). Demand and supply of feed ingredients for farmed fish and crustaceans: trends and prospects, FAO Fisheries and Aquaculture Technical Paper No. 564. Rome: Food and Agriculture Organization of the United Nations.

Taklemariam, A., Tessema, F., and Abayneh, T. (2015). Review on evaluation of safety of fish and fish products. International Journal of Fisheries and Aquatic Studies, 3, 111–117.

The ALS Association. (2018). The Ice Bucket Challenge Process. Retrieved March 12, 2018, from http://www.alsa.org/Figureht-als/ice-bucket-challenge.html

The Law Library of Congress, Global Legal Research Center. (2014, March). Restrictions on Genetically Modified Organisms. Retrieved January 22, 2017, from https://www.loc.gov/law/help/restrictions-on-gmos/restrictions-on-gmos.pdf

Torrissen, O., Jones, S., Asche, F., Guttormsen, A., Skilbrei, O., Nilsen, F., Horsberg, T.E., and Jackson, D. (2013). Salmon lice–impact on wild salmonids and salmon aquaculture. Journal of Fish Diseases, 36, 171–194.

Trushenski, J., and Bowzer, J. (2013). Having your omega 3 fatty acids and eating them, too: strategies to ensure and improve the long-chain polyunsaturated fatty acid content of farm-raised fish. In F. De Meester, R. Watson, and S. Zibadi, Sustainable Long Chain Omega-3 Fatty Acids in Cardiovascular and Mental Health (pp. 319–340). New York: Springer.

Trushenski, J., and DeKoster, T. (2017). Wild or farmed? Nutritional value of farmed versus wild white-fleshed fish is comparable. Retrieved from Global Aquaculture Advocate.

Trushenski, J., Schwarz, M., Bergman, A., Rombenso, A., and Delbos, B. (2012). DHA is essential, EPA appears largely expendable, in meeting the n – 3 long-chain polyunsaturated fatty acid requirements of juvenile cobia *Rachycentron canadum*. Aquaculture, 326, 81–89.

Trushenski, J., Blankenship, H., Bowker, J., Flagg, T., Hesse, J., Leber, K., MacKinlay, D.D., Maynard, D.J., Moffitt, C.M., Mudrak, V.A., Scribner, K.T., Stuewe, S.F., Sweka, J.A., Whelan, G.E., and Young-Dubovsky, C. (2015). Introduction to the HaMAR symposium: considerations for use of hatcheries and hatchery-origin fish. North American Journal of Aquaculture, 77:327–42.

Trushenski, J., Bowzer, J., Bergman, A., and Bowker, J. (2017). Developing rested harvest strategies for rainbow trout. North American Journal of Aquaculture, 36–52.

Trushenski, J.T., Whelan, G.E., and Bowker, J.D. (2018a). Why keep hatcheries? Weighing the economic cost and value of fish production for public use and public trust purposes. Fisheries, 43, 284–293.

Trushenski, J.T., Aardsma, M.P., Barry, K.J., Bowker, J.D., Jackson, C.J., Jakaitis, M., McClure, R.L., and Rombenso, A.N. (2018b). Oxytetracycline does not cause growth promotion in finfish. Journal of Animal Science, 96, 1667–1677.

Tukia, A. (2017, May 22). Thousands of dollars lost as salmon make slippery getaway. Newshub. Retrieved September 10, 2017, from www.newshub.co.nz/home/new-zealand/2017/05/thousands-of-dollars-lost-as-salmon-make-slippery-getaway.html

Turner, J. (2013). Genetically engineered crops and foods. In C. Bates, and J. Ciment, Global Social Issues: An Encyclopedia. London: Routledge.

Tveteras, S., Asche, F., Bellemare, M., Smith, M., Guttormsen, A., Lem, A., Lien, K., and Vannuccini, S. (2012). Fish is food – the FAO's fish price index. PLOS One, 7, e36731.

U.S., Control and Prevention. (2015, August 10). Heart Disease. Retrieved January 8, 2017, from Heart disease facts: https://www.cdc.gov/heart disease/facts.htm

U.S. Department of Health and Human Services/U.S. Department of Agriculture. (2015, December). Retrieved January 8, 2017, from Dietary Guidelines for Americans 2015–2020: https://health.gov/dietaryguidelines/2015/guidelines/

U.S. Environmental Protection Agency. (2011). National listing of fish advisories questions and answers 2011. Retrieved February 12, 2017, from https://www.epa.gov/fish-tech/national-listing-fish-advisories-questions-and-answers-2011

U.S. Environmental Protection Agency. (2017a, September 12). NPDES Aquaculture Permitting. Retrieved February 3, 2018, from National Pollutant Discharge Elimination System (NPDES): https://www.epa.gov/npdes/npdes-aquaculture-permitting#permit

U.S. Environmental Protection Agency. (2017b). Retrieved July 17, 2017, from Enforcement and Compliance History Online: https://echo.epa.gov/

U.S. Fish and Wildlife Service. (2012, January). The cost of invasive species. Retrieved May 6, 2017, from https://www.fws.gov/verobeach/PythonPDF/CostofInvasivesFactSheet.pdf

U.S. Fish and Wildlife Service. (2014, September 10). Common Carp (Cyprinus carpio) ecological risk screening summary. Retrieved May 6, 2017, from https://www.fws.gov/fisheries/ans/erss/highrisk/Cyprinus-carpio-WEB-09–10–2014.pdf

U.S. Fish and Wildlife Service. (2016, January 5). Aquatic Animal Drug Approval Partnership Program INAD Program. Retrieved November 6, 2016, from 17a-methyltestosterone INAD #11–236: https://www.fws.gov/fisheries/aadap/inads-available/medicated-feeds/17a-methyl testosterone/index.html

U.S. Fish and Wildlife Service. (2017). Retrieved April 30, 2017, from National Wild Fish Health Survey Database: National Wild Fish Health Survey Database

U.S. Food and Drug Administration. (1999a, January 14). Retrieved November 5, 2016, from http://www.fda.gov/downloads/animalvet erinary/products/approvedanimaldrugproducts/foiadrugsummaries/ucm116742.pdf

U.S. Food and Drug Administration. (1999b, April 5). Retrieved November 5, 2016, from http://www.fda.gov/downloads/AnimalVeterinary/Guid anceComplianceEnforcement/GuidanceforIndustry/UCM052375.pdf

U.S. Food and Drug Administration. (2005, October 24). Retrieved November 5, 2016, from http://www.fda.gov/downloads/Animal Veterinary/Products/ApprovedAnimalDrugProducts/FOIADrug Summaries/UCM051491.pdf

U.S. Food and Drug Administration. (2011). Enforcement Priorities for Drug Use in Aquaculture. Retrieved September 23, 2016, from http://www.fda.gov/downloads/AnimalVeterinary/GuidanceCompliance Enforcement/PoliciesProceduresManual/ucm046931.pdf

U.S. Food and Drug Administration. (2015a, July 7). Non-TDS Foods Analyzed for PCDD/PCDFs in 2001–2003. Retrieved March 11, 2017, from https://www.fda.gov/Food/FoodborneIllnessContaminants/Che micalContaminants/ucm077465.htm

U.S. Food and Drug Administration. (2015b, December 9). Dioxin analysis results/exposure estimates. Retrieved March 11, 2017, from https://www.fda.gov/Food/FoodborneIllnessContaminants/Chemical Contaminants/ucm077444.htm

U.S. Food and Drug Administration. (2015c, May). Retrieved November 5, 2016, from http://www.fda.gov/downloads/AnimalVeterinary/Guid anceComplianceEnforcement/GuidanceforIndustry/ucm052532.pdf

U.S. Food and Drug Administration. (2015d, June 3). Veterinary Feed Directive. Retrieved November 5, 2016, from Federal Register: https://www.federalregister.gov/documents/2015/06/03/2015–13393/veterinary-feed-directive

U.S. Food and Drug Administration. (2015e, December 21). AquAdvantage Salmon Fact Sheet. Retrieved January 22, 2017, from http://www.fda.gov/AnimalVeterinary/DevelopmentApprovalProcess/GeneticEngineering/GeneticallyEngineeredAnimals/ucm473238.htm

U.S. Food and Drug Administration. (2016, March 4). Approved Aquaculture Drugs. Retrieved November 5, 2016, from http://www.fda.gov/AnimalVeterinary/DevelopmentApprovalProcess/Aquaculture/ucm132954.htm

U.S. Food and Drug Administration. (2017a, January). Eating fish: what pregnant women and parents should know. Retrieved February 12, 2017, from http://www.fda.gov/Food/FoodborneIllnessContaminants/Metals/ucm393070.htm

U.S. Food and Drug Administration. (2017b, February 3). Dioxin. Retrieved March 11, 2017, from https://www.fda.gov/AnimalVeterinary/Products/AnimalFoodFeeds/ucm050430.htm

U.S. Food and Drug Administration. (2017c, November 6). The ins and outs of extra-label drug use in animals: a resource for veterinarians. Retrieved July 30, 2018 from https://www.fda.gov/animalveterinary/resourcesforyou/ucm380135.htm

U.S. Food and Drug Administration. (2017d). FDA's strategy on Antimicrobial Resistance - Questions and Answers. Retrieved July 30, 2018 from https://www.fda.gov/animalveterinary/guidancecomplianceenforcement/guidanceforindustry/ucm216939.htm.

U.S. Food and Drug Administration. (n.d.). Animal Drugs @ FDA. Retrieved November 5, 2016, from https://animaldrugsatfda.fda.gov/adafda/views/#/search

U.S. Geological Survey. (2016a, December 2). How much water is there on, in, and above Earth? Retrieved July 8, 2017, from The USGS Water Science School: https://water.usgs.gov/edu/earthhowmuch.html

U.S. Geological Survey. (2016b, December 9). Estimated use of water in the United States county-level data for 2010. Retrieved July 8, 2017, from Water Use in the United States: https://water.usgs.gov/watuse/data/2010/index.html

U.S. Government. (2016). Code of Federal Regulations. Title 21 Food and Drugs, Chapter 9 Federal Food, Drug, and Cosmetic Act, Subchapter

II Definitions. Retrieved September 23, 2016, from http://www.ecfr. gov/cgi-bin/text-idx?SID=3ee286332416f26a91d9e6d786a604aband mc=trueandtpl=/ecfrbrowse/Title21/21tab_02.tpl

U.S. National Oceanic and Atmospheric Administration. (2016, August). Fisheries of the United States, 2015. Retrieved January 8, 2017, from Commercial Fisheries Statistics: https://www.st.nmfs.noaa.gov/ commercial-fisheries/fus/fus15/index

Uglem, I., Bjorn, P., Dale, T., Kerwatch, S., Okland, F., Nilsen, R., Aas, K., Fleming, I., and McKinley, R. (2008). Movements and spatiotemporal distribution of escaped farmed and local wild Atlantic cod (*Gadus morhua L.*). Aquaculture Research, 39, 158–170.

UNESCO. (2015). World Water Development Report 2015. Paris, France: United Nations Educational, Scientific, and Cultural Organization.

United Nations. (n.d.). Facts and Figures. Retrieved July 8, 2017, from United National International Year of Water Cooperation: www. un.org/en/events/worldwateryear/factsfigures.shtml

United Nations Statistics Division. (2015). Retrieved January 29, 2017, from The World's Women 2015: http://unstats.un.org/unsd/gender/ worldswomen.html

US National Institutes for Health. (2016). Eunice Kennedy Shriver National Institute of Child Health and Human Development. Retrieved June 18, 2016, from Zebrafish Core: https://www.nichd.nih. gov/about/org/dir/osd/cf/zc/Pages/overview.aspx

van Senten, J., and Engle, C. (2017). The costs of regulation on US baitfish and sportfish producers. Journal of the World Aquaculture Society, 48, 503–517.

Verbeke, W., Sioen, I., Brunso, K., De Henauw, S., and Van Camp, J. (2007). Consumer perception versus scientific evidence of farmed and wild fish: exploratory insights from Belgium. Aquaculture International, 15, 121–136.

Vincent, D. (2004) Why Montana went wild. Montana Outdoors, Montana Fish Wildlife and Parks. Retrieved from: http://fwp.mt.gov/ mtoutdoors/HTML/articles/2004/DickVincent.htm

Wagner, E., and Oplinger, R. (2013, January). Review of sterile fish production using hybrid crosses or ploidy manipulation. Retrieved January 22, 2017, from Fisheries Experiment Station: https://wildlife.utah.gov/ fes/pdf/sterile_fish_production.pdf

Walsh, B. (2011, July 7). The end of the line. Time. Retrieved July 1, 2017, from content.time.com/time/health/article/0,8599,2081796,00.html

Watson, R., and Pauly, D. (2001). Systematic distortion in world fisheries catch trends. Nature, 424, 534–536.

Welch, A. (2015). Farming in the commons, fishing in the Congress, and U.S. aquaculture in the 21st Century (dissertation). Coral Gables, USA: University of Miami. Retrieved July 1, 2017, from http://scholarlyrepository.miami.edu/oa_dissertations/1486

Windsor, M. (1971). Fish meal, Torry Advisory Note No. 49. Torry, Aberdeen, UK: Department of Trade and Industry, Torry Research Station. Retrieved June 23, 2016, from http://www.fao.org/wairdocs/tan/x5926e/x5926e00.htm#Contents

World Bank. (2007). Changing the face of the waters. Washington, D.C.: World Bank.

World Bank. (2016a). Commodity prices, Fishmeal. Retrieved July 9, 2016, from Index Mundi: www.indexmundi.com

World Bank. (2016b). Poverty and shared prosperity 2016: taking on inequality. Washington, D.C.: World Bank.

World Bank. (2017a, January 3). Retrieved January 14, 2017, from World Bank Open Data: http://data.worldbank.org/

World Bank. (2017b, March). World Bank list of economies. Geneva, Switzerland. Retrieved May 29, 2017, from databank.worldbank.org/data/download/site-content/CLASS.xls

World Bank/Food and Agriculture Organization of the United Nations/International Fund for Agricultural Development. (2009). Gender in agriculture sourcebook. Washington, D.C.: World Bank.

World Health Organization. (2015). WHO estimates of the global burden of foodborne diseases. Geneva, Switzerland: World Health Organization.

World Organization for Animal Health. (2016). Manual of diagnostic tests for aquatic animals. Paris, France: World Organization for Animal Health. Retrieved May 6, 2017, from www.oie.int/international-standardsetting/aquatic-manual/access-online/

World Resources Institute. (2005). Ecosystems and human well-being – millennium ecosystem assessment. Washington, DC: Island Press.

Woynarovich, A., Moth-Poulsen, T., and Peteri, A. (2010). Carp polyculture in Central and Eastern Europe, the Caucasus and Central Asia. Rome, Italy: Food and Agriculture Organization of the United Nations.

Wright, J. (2016, October 17). Omega-3 levels fall in farmed salmon but it's still a top source. Retrieved January 8, 2017, from Global

Aquaculture Advocate: http://advocate.gaalliance.org/omega-3s-levels-fall-in-farmed-salmon-but-its-still-a-top-source/

Yorktown Technologies, L.P. (2017). Glofish, experience the Glo! Retrieved January 16, 2017, from Glofish Science: http://www.glofish.com/about/glofish-science/

Young, N., and Matthews, R. (2010). The aquaculture controversy in Canada: activism, policy, and contested science. Vancouver: UBC Press.

Zhu, Y., and Chu, J. (2013). A study on aquaculture industry regulations and policies of globally representative states. Chinese Journal of Population Resources and Environment, 11, 268–275.

Index